Immunologically Active Peptides

Developments in Molecular and Cellular Biochemistry

Volume 2

V. A. NAJJAR, *series editor*

Already published:
1. V. A. Najjar, ed., The Biological Effects of Glutamic Acid and Its Derivatives, 1981. ISBN 90-6193-841-4.

series ISBN 90-6193-898-8.

SPRINGER-SCIENCE+BUSINESS MEDIA, B.V. 1981

Immunologically Active Peptides

Edited by

V. A. NAJJAR

Reprinted from
Molecular and Cellular Biochemistry, Vol. 41, 1981

SPRINGER-SCIENCE+BUSINESS MEDIA, B.V. 1981

Distributors:

for the United States and Canada

Kluwer Boston, Inc.
190 Old Derby Street
Hingham, MA 02043
USA

for all other countries

Kluwer Academic Publishers Group
Distribution Center
P.O. Box 322
3300 AH Dordrecht
The Netherlands

Library of Congress Cataloging in Publication Data

Main entry under title:

Immunologically active peptides.

    (Developments in molecular and cellular biochemistry; v. 2)
    'Reprinted from Molecular and cellular biochemistry, vol. 41, 1981.'
    Includes index.
    1. Peptides – Physiological effect – Addresses, essays, lectures.
    2. Immunochemistry – Addresses, essays, lectures. I. Najjar, V. A.
(Victor A.), 1914–. II. Molecular and cellular biochemistry. III. Series.
QP552. P4145 1981   616.07′9   81-20732
                  AACR2

ISBN 978-94-009-8032-7   ISBN 978-94-009-8030-3 (eBook)
DOI 10.1007/978-94-009-8030-3

Contents

Introduction to Developments in molecular and cellular biochemistry

Molecular and Cellular Biochemistry is an international journal that covers a wide range of biophysical, biochemical and cellular research. This type of coverage is intended to acquaint the reader with the several parameters of biological research that are relevant to various fields of interest. Unlike highly specialized journals, it does not bring into focus a particular field of investigation on a monthly basis. Accordingly, it has been decided to supplement its present wide scope by periodic presentations of a restricted area of research in the form of book-length volumes. These volumes will also be published in hardcover as a book series entitled, *Developments in Molecular and Cellular Biochemistry*. Each volume will focus on an active topic of interest which will be covered in depth. It will encompass a series of contributions that deal exclusively with one single well-defined subject.

The present volume, *Immunologically Active Peptides,* is the second one in the book series. The first one deals extensively with *The Biological Effects of Glutamatic Acid and Its Derivatives.*

It is the editor's hope that *Molecular and Cellular Biochemistry* will fulfill its intended role of service to the international community of biological scientists.

Immunologically active peptides

No area of science, in this decade, has witnessed a greater expansion than the science of immunology, both cellular and molecular. The mere classification of the various types of lymphocytes, both immunochemically and biologically, continues unabated. The phagocytic cells have bounced back from relative obscurity to well-deserved prominence. The macrophage is a world of its own. The so-called 'scavenger', a grave insult to a noble cell, has now become the hub of immunology as it sits at the convergence of cellular and numoral immunology. It is at once a sequesterer and killer of invading bacteria and syngeneic cancerous cells, as well as, the processor of the antigen for anti-body response. Such a cell, one would think, is entitled, in its own right, to a natural activator that regulates its function. Various factors have one or more qualifications for this role. Several articles in this volume bear on this point. Little need be said for the various lymphokines, thymosin, thymic serum factor, etc., as these polypeptides have already taken a well-earned place in the realm of immunology. Several articles deal with this aspect of biology and function.

Victor A. Najjar,
Editor in chief

Tuftsin, a natural activator of phagocytic functions including tumoricidal activity

Victor A. Najjar, Danuta Konopinska, Manas K. Chaudhuri, Donald E. Schmidt* and Lisa Linehan
*Division of Protein Chemistry, Tufts University School of Medicine, Boston, MA 02111, * Rainin Instruments Co., Inc., Woburn, MA 01801, U.S.A.*

Summary

Some of the properties of the tetrapeptide tuftsin, Thr-Lys-Pro-Arg, are discussed. We describe three phases of tuftsin activation of the macrophage. Tuftsinyltuftsin, the octapeptide Thr-Lys-Pro-Arg-Thr-Lys-Pro-Arg, was synthesized with a view of minimizing the formation of Lys-Pro-Arg, from tuftsin by tissue aminopeptidases. The tripeptide is a tuftsin inhibitor. The octapeptide proved to be quite effective in prolonging the life of syngeneic mice injected with L1210 leukemia cells. Its effect in our laboratory, was considerably better than we could obtain with tuftsin. A simple method for purifying tuftsin by high performance liquid chromatography is described using 0.75% trifluoroacetic acid in water.

The tuftsin sequence Thr-Lys-Pro-Arg is present in P12 protein of Rausher murine leukemia virus. A close analog Thr-Arg-Pro-Lys appears in yet another virus protein the haemagglutinin of influenza virus. A second close analog Thr-Arg-Pro-Arg forms the penultimate carboxyterminal of a pancreatic polypeptide found in human and several animals.

Introduction

Over four dozen compounds have so far been reported to activate phagocytic cells; principally the macrophage. The mechanism of activation is a highly complex process that involves the stimulation of several functions of the phagocyte. These are primarily phagocytosis, kinesis, immunogenic activity hexose monophosphate shunt, bactericidal activity and most importantly tumoricidal activity.

It is hardly possible that nature would devise such a highly elaborate and complex system only to be triggered by compounds external to the body such as bacteria and bacterial products. Nature would do better. It would devise a natural intrinsic compound(s) primarily designed for the sole purpose of regulating the physiological functions of phagocytic cells. Among these is a compound which was discovered a few years ago in human serum is tuftsin (1–5) a tetrapeptide, Thr-Lys-Pro-Arg.

A good deal of novel and unique observations preceded the identification of tuftsin (6, 7). These findings gave rise to the conviction that it is a true biological entity with the assigned function of activating phagocytic cells: (a) Tuftsin is a part of a specific carrier γG_1 cytophilic γ-globulin, leukokinin (8). (b) The carrier leukokinin circulates through the spleen where it is nicked by an enzyme, tuftsin-endocarboxypeptidase, between residues 292 and 293 (Arg-Glu) of the heavy chain. This nicked molecule, leukokinin-S (2), then binds to specific receptors on the outer membrane of the phagocyte where the enzyme leukokininase cleaves the tetrapeptide off the parent molecule between residues 288 and 289 of the heavy chain (Lys-Thr) to release active tuftsin (2). Thus the phagocytic cell plays a unique role in releasing its own activator tuftsin. (c) Finally, a mutation affecting tuftsin has been identified. It results in a specific human syndrome with increased frequency of severe infections. Seventeen such mutants have been identified in the

Molecular and Cellular Biochemistry 41, 3–12 (1981). 0300-8177/81/0041-0003/$02.00.
© 1981, Martinus Nijhoff/Dr W. Junk Publishers,

United States of America (9–11) and two in Japan (12). (d) Another type of tuftsin deficiency is encountered after removal of the spleen (8–11, 13, 14). Here the tetrapeptide fails to be nicked at Arg-Glu and consequently remains bound to leukokinin and as such is inactive. This type of intact leukokinin still binds as tightly to its cell receptors as leukokinin-S which is already nicked at Arg-Glu (Table 1). The table shows that in low ionic strength isotonic buffers, the binding to specific cell receptors is equally strong for leukokinin-S and for unnicked, intact and inactive leukokinin. This indicates that bound leukokinin of either type displays tight and continuous association with the cell surface. Consequently, the receptor site is distinct and separate from the site of leukokininase activity.

The foregoing findings which are detailed in several reviews (2, 3, 15) support the concept that tuftsin is a real biological entity with a specific defensive function exerted on phagocytic cells, which are the only cells that possess specific tuftsin receptors (13). The importance of an activator like tuftsin is reflected in the fact that nature has

preserved its fundamental structure. It is present in some analogous form in all animals studied. It has been isolated in pure form from man and dog (2). In the latter, the carboxy-terminal residue, arginine, of human tuftsin is replaced by lysine. The resulting analog exhibits comparable biological activity. In the rabbit, immunochemical studies indicate that the second, third and fourth residues from the amino terminal are the same as in man (13, 14). Furthermore, in mice γ-G1, MOPC-21 a glutamine residue replaces lysine, the second residue of tuftsin (16).

The activation of the phagocytes

The functional expression of tuftsin is exerted in three phases:

Phase I is expressed in rapid stimulation of the phagocytic and pinocytic activity of the neutrophilic granulocyte and the macrophage. There is also a concomitant stimulation of the hexose monophosphate pathway.

Phase II is a more gradual process that extends

Table 1. Serum was obtained from dog #25 a week before splenectomy (leukokinin-S) and three months after splenectomy. Leukokinin was prepared from serum γ-globulin by fractionation (Fraction IV) on phosphocellulose columns (32). Phagocytosis assay (7–9) was carried out with dog buffy coat cells in isotonic buffered sucrose using leukokinin prepared from serum obtained before splenectomy, experiments 1–4, and after splenectomy, experiments 5–8. Assays were further carried out with a mixture of the two types of leukokinin but with a definite order of addition. In the first addition, cells were allowed to bind added leukokinin for 5 min after which the second addition was made along with target particles *Staphylococcus aureus*. Note that phagocytosis stimulation was determined by the type of leukokinin used in the first addition showing tight binding to membrane receptors. Phagocytic index represents the number of cells containing bacteria/100 phagocytic cells counted usually 400–500 cells were counted.

Experiment number	Order of addition of leukokinin prepared from serum obtained before (leukokinin-S) or after splenectomy (Dog #25)		Leukokinin each addition (μg)	Phagocytosis index (%)
	1st addition	2nd addition		
1.	Before splenectomy	None	0	18
2.	Before splenectomy	None	50	35
3.	Before splenectomy	None	100	42
4.	Before splenectomy	None	200	41
5.	After splenectomy	None	0	16
6.	After splenectomy	None	50	18
7.	After splenectomy	None	100	21
8.	After splenectomy	None	200	19
9.	Before splenectomy	After splenectomy	100	40
10.	Before splenectomy	After splenectomy	200	45
11.	After splenectomy	Before splenectomy	100	22
12.	After splenectomy	Before splenectomy	200	19

for a period of a few hours during which there is a strong stimulation of motility. At the same time the engulfed particle or antigen is degraded or processed for presentation to the antibody forming cell.

Phase III is a much slower process than the first two and requires several days. This phase or in combination with other phases, in one way or another, works towards the maturation and activation of the large spreading phagocytic cell to augment its bactericidal activity and to render it highly cytocidal towards syngeneic tumor cells. Details of the three phases follow:

Phase I:

(a) *Increased phagocytic activity.* Increased phagocytic and pinocytic activity is an immediate and highly specific effect of tuftsin. It can be measured within a few minutes after the exposure of blood granulocytes or tissue macrophages to low concentrations of tuftsin. The concentration that results in half maximal phagocytic activity (Km) of either phagocytic cell is about 100 ηM (0.05 μg/ml) (3, 13–15). Tuftsin is incapable of stimulating pinocytosis of $3T_{12}$, $3T_6$ and L1210 cells (15) as well as HeLa cells (Constantopoulos, A. and Najjar, V. A., unpublished).

(b) *Increased stimulation of the hexosemonophosphate shunt.* It has been known for sometime that phagocytosis is accompanied by stimulation of the hexose monophosphate shunt with its attendant increase in hydrogen peroxide, superoxide, hydroxyl radicals and singlet oxygen (17). The high level of these active compounds is assumed to result in the halogenation of the ingested particle or a direct effect of these compounds on the target particle with its consequent destruction and elimination, be it a bacterium, a dead cell or an abnormal cell (15, 18).

Stimulation of the metabolic shunt by tuftsin, as measured by the reduction of the nitrotetrazolium dye, was shown by Spirer *et al.,* (19). It follows that under these conditions there would be expected an increased rate of killing of ingested bacteria. This was indeed the case. Martinez *et al.,* (20) showed conclusively that four different species of bacteria were killed more rapidly after ingestion by tuftsin activated macrophages than by control cells not exposed to tuftsin. This bactericidal stimulation paralleled the increased blood clearing of these bacteria following tuftsin injection of 10–20 mg/kilo of body weight.

Phase II:

(a) *Stimulation of motility of blood granulocytes.* In contrast to phagocytic stimulation, the increase rate of motility requires considerably higher concentrations of tuftsin, approximately fiftyfold. At 5 μM there was definite increase in motility, approximately a third of maximal. However at 25 μM, maximum stimulation was attained. Thus at a concentration that yields a half maximal phagocytic effect 0.10 μM, no stimulation of motility could be discerned. This is of particular importance since the concentration available to the blood neutrophil in human serum is approximately 0.3 μM (14). Consequently, it is difficult to assign a definitive biological role for tuftsin stimulation of motility in view of the high concentration required unless the effective concentration is much greater. This could well be the case inasmuch as tuftsin carrier leukokinin is bound to cell receptors in saturating amounts at normal serum levels (Table 1). Furthermore, tuftsin is released on site.

(b) *Stimulation of the immunogenic function of the macrophage.* This phase occurs after prolonged exposure of the macrophage to the antigen during which a T helper cell, in the blast stage, would interact with the antigen processing macrophage. The presence of tuftsin during this time stimulates considerably the little understood phenomenon of processing of the antigen for maximal immunogenic stimulation. Mouse peritoneal macrophages were exposed to thymus dependent antigen with and without tuftsin (21). After a few hours of incubation, the washed macrophages were then incubated with nonadherent splenic cells. These were later irradiated and injected into syngeneic mice. After several days, the lymphocytes of the draining lymphnodes were shown to take up

[^{3}H]-thymidine over seven times the uptake of similar lymphocytes obtained from the corresponding controls where no tuftsin was used during antigen processing. Maximal stimulation was obtained at tuftsin concentration of approximately 5 μM. Concentrations on either side of the optimum result in a diminished effect (21). It is our opinion that this regulatory pattern is modulated by the relative concentrations of tuftsin and the tripeptide Lys-Pro-Arg, a product of the action of an aminopeptidase on tuftsin and which is a good inhibitor of tuftsin stimulation of phagocytosis (13).

The demonstration at the Weizmann Institute of the immunogenic stimulation by tuftsin which is a distinct part of the repertoir of the macrophage may be translated into a direct augmentation of antibody production by antibody forming cells. This indeed proved to be the case as demonstrated clearly by Florentin et al., (22). They showed that tuftsin injected into mice at a precise number of days before antigen administration, increased antibody response significantly to T dependent and independent antigens as well as augmented the antibody-dependent cell mediated cytotoxicity.

Phase III:

(a) *Stimulation of maturation and differentiation of the macrophage.* There have been of late, several instances where the effect of tuftsin administration to an animal requires several days to reach full fruition. Thus the experiments reported by Florentin et al., (22) clearly show that, following tuftsin administration to mice, several days must elapse before antigen challenge at which time the recipient animal is fully prepared to mount a maximal immune response. For T dependent antigens, seven days must elapse before antigen injection and only 1-3 days for T independent antigens, while antibody-dependent cell mediated cytotoxicity require seven days for its fullest expression. These findings parallel the results obtained by Nishioka (23) showing that 5-7 days after the intraperitoneal injections of tuftsin in mice, there was a considerable augmentation of the number of mature spreading intraperitoneal macrophages.

(b) *Stimulation of the tumoristatic and tumoricidal activity of the macrophage by tuftsin and tuftsinyltuftsin.* A good deal of work has been done to establish the fact that mononuclear cells including tissue macrophages, blood monocytes and granulocytes can exert antineoplastic properties (24, 25). The question naturally arises as to whether tuftsin can augment such properties inasmuch as it is the natural activator of these cells (15).

Florentin et al., (22) and Nishioka (23) showed that after tuftsin-treatment, mice developed a higher level of cytostatic and cytocidal activity of their macrophages. Furthermore, there was also an enhancement of antibody-dependent cell-mediated cytotoxicity (22). This activation of the macrophage by tuftsin resulted in an in vivo augmentation of mice survival following injections with L1210 cells and suppression of the growth of injected melanoma cells (23).

Another successful demonstration of the antitumor activity of tuftsin in vivo has recently been reported by Catane et al., (26). They used 3-methyl-cholanthrene induced transplantable fibrosarcoma injected intraperitoneally, in C3H mice. This was uniformly lethal in control animals with a mean survival time of 21 days. However, with the intraperitoneal injection of 10-500 μg/kg of body weight, the mean survival time was 39 days (p < 0.001) and 20% of the mice survived beyond 80 days.

Our laboratory has been similarly engaged in a parallel effort for the study of the effect of tuftsin on the survival of L1210 infected mice. We obtained increased survival of treated mice but not of sufficient magnitude. It was our opinion that with the most active preparation of tuftsin, the formation of the tripeptide Lys-Pro-Arg, a natural inhibitor of tuftsin (13), might explain the limited effect obtained. There are active cytosol and membrane aminopeptidases that would yield this inhibitor peptide from tuftsin (15). Consequently, we synthesized several analogs in order to evaluate this antitumor activity with the primary purpose of minimizing or eliminating the formation or the accumulation of Lys-Pro-Arg. Among these are Ala-Lys-tuftsin, acetyl-Ala-Lys-tuftsin, Ala-Lys-tuftsin-Glu-Ala-Ala-Ala, [Glu^2]-tuftsin and the octopeptide tuftsinyltuftsin (Thr-Lys-Pro-Arg-Thr-Lys-Pro-Arg) (27). In each case, it is expected that the tetrapeptide would be released slowly and little if any Lys-Pro-Arg would accumulate. In this connection, a novel tuftsin analog in which threonine is cyclized (28), [O = C$\widehat{\vee}$Thr¹]-tuftsin, is eminently suitable for this purpose since it would be more stable to aminopeptidase degradation and would give rise to little if any of the inhibitor Lys-Pro-Arg. Twenty five DBA/2 mice were injected each with 20 µg of the octapeptide intraperitoneally on days -7, -4 and 0. On the latter day, L1210 cells were also injected intraperitoneally in experimental animals and solvent buffer alone in 25 control mice (27).

Fig. 1 illustrates the considerable increase in the survival of mice treated with tuftsinyltuftsin (p < 0.001). Eventually, these animals died presumably because a sufficient number of

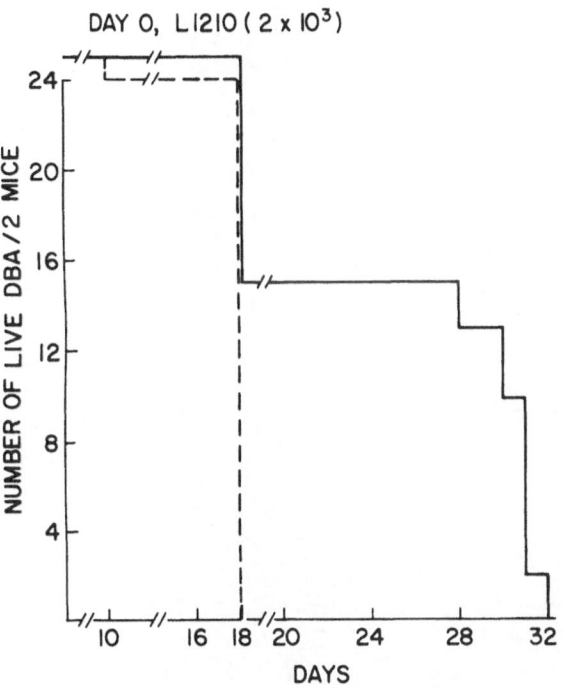

DAY 0, L1210 (2 x 10³)

Fig. 1. 25 DBA/2 female mice were injected intraperitoneally with 20 µg each of tuftsinyltuftsin on day -7, -4 and 0, at 0 time. Each also received 2 × 10³ L1210 cells which were frozen immediately after harvesting from the peritoneal cavity of infected mice. Twenty-five control mice were injected with saline. The animals were observed daily and deaths recorded.

Control animals were all dead by day 18, whereas, treated mice survived 32 days.

L1210 cells were not destroyed by the octapeptide activated peritoneal macrophages inasmuch as the state of macrophage activation is finite. In a similar experiment (not shown), a high percentage of treated animals this time survived for over three months and showed no evidence of disease, while none of the control mice survived beyond 42 days. The lungs, liver, spleen and pancreas of the untreated control animals were heavily infiltrated with L1210 cells. In contrast the tissues of surviving treated animals were not infiltrated. Fig. 2 shows representative tissues from untreated control and surviving treated mice. This represents one of several experiments (27), (P < 0.001) where tuftsinyltuftsin was used on DBA/2 mice 11–13 weeks old. Currently, we are testing the effect of continued weekly injections of tuftsin and tuftsinyltuftsin after injection with L1210 cells.

Tuftsin receptor sites and anti-tuftsin antibody

From previous theoretical considerations (29), it would be predictable that the antibody site directed against tuftsin would mimic closely the receptor sites to which tuftsin binds. This parallels recent findings that antibody against insulin receptors has insulin-like properties (30) and confirms our earlier report (31) where antibody was generated in rabbits against horse antibody to lecithinase. This latter antibody reacts directly at the active site of the enzyme (32). The rabbit antibody against the horse antibody paratope surface would be a close copy of the enzyme site. In fact, it was shown to portray enzyme like properties in that it bound the substrate lecithin (31). Normal serum and several antisera to various antigens did not bind lecithin (31).

If indeed, antibody to tuftsin mimics tuftsin receptors, then strong inhibitor analogs to tuftsin activity would be expected to bind more strongly to the antibody binding site than tuftsin (27). Thr-Lys-Pro-Pro-Arg completely inhibits phagocytic stimulation by tuftsin at one third the concentration of the tetrapeptide (5) and is a strong competitive inhibitor for tuftsin binding to its antibody (27). The same holds true for Lys-Pro-Arg which is also a good inhibitor of tuftsin and binds more strongly to anti-tuftsin antibody (14). Similarly, the mutant peptide, tentatively identified in one tuftsin deficiency patient, Thr-Glu-Pro-Arg, shows complete

8

inhibition of tuftsin stimulation of phagocytosis at one half the concentration of tuftsin (9–11). At the same time, it also exhibits stronger binding to tuftsin antibody than tuftsin itself (27). Finally, tuftsinyltuftsin which is biologically active in vivo presumably, because it yields tuftsin through tryptic like cleavage, is inhibitory to phagocytic stimulation and also binds to the specific antibody more strongly than tuftsin. The foregoing emphasizes the similarity of antibody binding to tuftsin and the

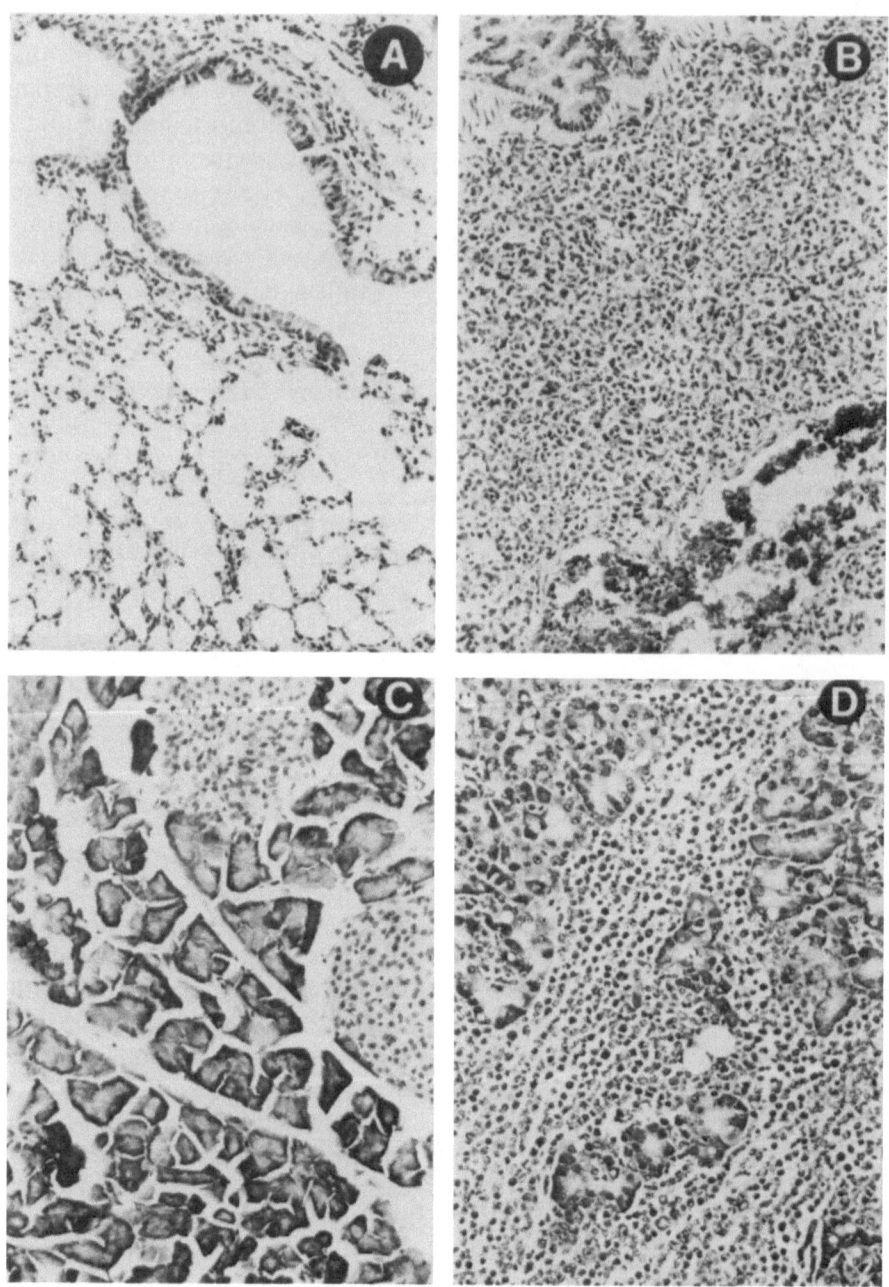

Fig. 2. A. Surviving treated mouse lung. Hematoxylin and eosin X 250. Normal histologic appearance. B. Untreated control mouse lung. Hematoxylin and eosin X 250. The architecture is obliterated by massive infiltraton by L1210 cells. C. Surviving treated mouse pancreas. Hematoxylin and eosin X 250. Normal acinar and islet tissue. D. Untreated control mouse pancreas. Hematoxylin and eosin X 250. The interstitium of the acinar tissue is heavily infiltrated by L1210 cells.

binding of tuftsin to receptors on the plasma membrane of the phagocyte. It also brings out the possible error that can arise in radioimmunoassay methodology if applied to subjects that may represent a tuftsin mutation.

The occurence of tuftsin, Thr-Lys-Pro-Arg and its analogs, Thr-Arg-Pro-Lys and Thr-Arg-Pro-Arg in other proteins

On the assumption that all 20 amino acid appear in protein with equal frequency, the occurence then of tuftsin sequence by simple chance would be 20^4 (1:160 000). Since tuftsin is a part of the constant Fc region of the $G_1\gamma$-globulin molecule it is reasonable to expect its presence in a variety of subfractions (33) that differ only in the variable region of the Fab segment. This proved to be the case (2). However, tuftsin sequence Thr-Lys-Pro-Arg, does occur in residues 9–12 of P12 protein of Raucher murine leukemia virus (34). It appears that it is not a mere bystander in that molecule but signals a biological function for the carrier P12 protein. Treatment of cells infected with this virus in vitro with tuftsin resulted in a threefold increase in virus-associated reverse transcriptase. More impressive was a very considerable increase in virus induced membrane budding (35). Similar results were obtained by Suk and Long (36) with Kirsten Sarcoma virus grown on mouse cells. They showed that tuftsin at 0.01–100 μg/ml induces endogenous virus within 3–4 h. Actinomycin D blocked this process suggesting that RNA synthesis is required for the induction process.

The close analog Thr-Arg-Pro-Lys, where Lys and Arg exchange places in the tuftsin sequence, is found at residues 214–217 in yet another virus protein; the influenza haemagglutinin JAP HA (H2, ″57) (37). Its function in this protein remains to be elucidated although a likely role in haemagglutination is an attractive possibility because of its highly polar character.

In addition to this, there is another analog, Thr-Arg-Pro-Arg, in which the arginine residue replaces lysine the second residue in tuftsin sequence. This tuftsin analog appears in the biologically active pancreatic polypeptide at residues 32–35. This places this tetrapeptide at the extreme carboxy-terminal end next to the terminal residue tyrosinamide (38). The pancreatic polypeptide is presumed

to have gastrointestinal functions and a possible satiety factor. It is secreted after a protein meal and leads to a decrease in weight of certain genetically diabetic mice (38, 40). This tetrapeptide is present in all analogous pancreatic peptides of human, bovine, ovine, porcine and canine origin.

If must be emphasized that both Thr-Arg-Pro-Arg (41) and Thr-Arg-Pro-Lys (41, 42) are as active as tuftsin Thr-Lys-Pro-Arg.

Chipens and his colleagues (43) reported the synthesis of tuftsin and several analogs including circular analogs. They extended the spectrum of activity of tuftsin to include myotropic activity on the isolated rat ascending colon. The response amounted to 20% of that shown by angiotensin. They also made note of the fact that human heavy chain (EU) contains the sequence Pro-Lys-Pro-Arg (245–248), as well as Gly-Gln-Pro-Arg (341–344) termed rigin. The latter was synthesized and shown to have the same phagocytic activity as tuftsin (44). Conformational analysis showed a great deal of similarity between tuftsin and rigin.

The purity of tuftsin preparations

We have stressed on numerous occasions (3, 15) the importance of maximum purity of tuftsin preparations. One of the earliest attempts at making analogs showed clearly that many of these are inhibitory to tuftsin activity. It could well be that inconsistent results, in all parameters of tuftsin activity, could be due to the presence of inhibitory impurities of variable amounts and composition. Some of the possible analog impurities stem from the impurities present in the protected and activated amino acids that are purchased. For example, an incompletely protected amino acid preparation will definitely yield double coupling, one on the recipient residue and another on the incoming residue itself. Similarly, branched analog impurities can arise if functions other than the alpha amino group are not properly protected. In the case of tuftsin this is most likely to be an inhibitory analog and would render tuftsin inactive even if the impurity appears to be a small percent of the final product. It is remarkable how many tuftsin preparations are inactive with the phagocytosis assay only to become quite active upon purification. Consequently, a further purification on high performance liquid chromatography is necessary and

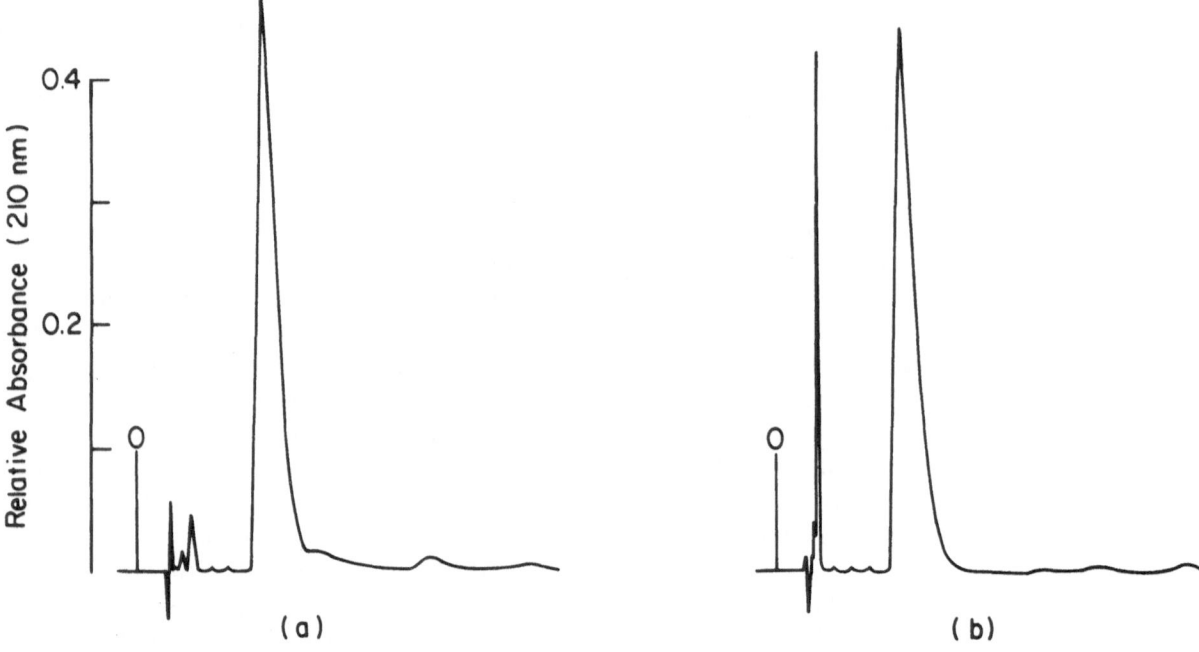

Fig. 3. Chromatograms on DuPont 850 HPLC (a) tuftsin 250 μg unpurified and synthesized in this laboratory by the solid phase method. (b) tuftsin 250 μg purchased from an American supplier. The main peak in each case is tuftsin. Column, DuPont ZORBAX TMS 9.4 mm × 25 cm. Absorbance Units Full Scall 0.32 (∞ denotes 5.12). Mobile phase 0.075% trifluoroacetic acid in water, flow rate 8 ml/min, chart rate 0.5 cm/min, wave length 210 ηm, T_{22}.

almost always reveals impurities not separated by previous methods (5). Fig. 3 represents two preparations of tuftsin with evidence of presumed purity in several systems of thin layer chromatography. Preparation (a) is synthetic tuftsin that was synthesized in our laboratory. Preparation (b) was purchased from an American supplier.

Concluding remarks

Tuftsin has been shown to possess a wide variety of properties all of which could be the result of activation of the macrophage and the blood granulocytes. Among these are increased phagocytosis and pinocytosis, stimulation of cell motility, augmentation of the immunogenic response with increased formation of antibody to T-dependent and independent antigens as well as the stimulation of antibody-dependent cell-mediated cytotoxicity. Tuftsin activation of the macrophage also results in increased cytostatic and cytotoxic activities toward bacteria and syngeneic tumor cells. The prolongation of the survival of mice injected with tumor

gives promise that this tetrapeptide, of human origin, may be a good candidate for additional human trials. Tuftsinyltuftsin appears to be more effective in its tumoricidal activity probably because of minimal or no accumulation of the inhibitory tripeptide Lys-Pro-Arg.

Epilogue

Inasmuch as it is the policy of this journal to encourage the disclosure of the circumstances that led to a particular discovery, it may be of some interest to do so for the tetrapeptide tuftsin. The discovery of this specific phagocyte activator was not the product of intuitive thinking, teleological considerations or foresight. It was accidental.

For sometime, our laboratory was engaged in the study of the physiological role of cytophilic γ-globulin. We had already shown that specific erythrophilic γ-globulin is essential for the survival of the red cell and leukophilic γ-globulin, leukokinin, was necessary for the full expression of the phagocytic activity of the blood granulocyte. It was during a study of the unique kinetics of the latter by one of us (V. A. Najjar) that a fortunate interruption of the experiment took place. All along, we had regarded the kinetics of leukokinin stimulation of phagocytosis as unusual (2) because the activation of the cell was short lived. After an initial stimulation by leukokinin, the rate of phagocytosis levels off early on and further addition of leukokinin results in a second stimulation which also levels off (2). It appeared that something other than the target particle substrate was being used up in the process and the question arose as to whether leukokinin was being inactivated. However, we were dealing with a cell and not an enzyme. Thus this anomaly was credited to the vagaries of cell behavior and all concerns in that regard were diffused.

We proceeded with other studies on the mechanism of leukokinin action. Under the usual procedure then employed for phagocytosis studies, a preparation of granulocytes was mixed with leukokinin at 37° to be followed by the addition of the bacteria to start the reaction. It was just before the addition of the bacteria that the experimentor was interrupted. It was quite some time before the experiment was continued. During that time the phagocytic cells were interacting with leukokinin. The experiment was resumed by the addition of target bacteria. For the first time the experiment had failed and there was no phagocytic stimulation. However, good stimulation was obtained in the same reaction mixture when another aliquot of leukokinin was added to these cells. After further experimentation, it became clear that the stimulatory effect of leukokinin was indeed short lived and that it had become inactivated in the process. Such spent leukokinin was incapable of stimulating a fresh preparation of granulocytes even though its overall biochemical and physical characteristics had not in the least been altered (2). Clearly it has lost a small fragment. That fragment was isolated, sequenced and synthesized, Thr-Lys-Pro-Arg, tuftsin.

Acknowledgement

We are very grateful to Dr. Eugene A. Foster, Department of Pathology, Tufts University School of Medicine who was kind enough to examine the tissues from the mice in the experiments with L1210 cells and prepare the photomicrographs for Fig. 2.

This work was supported by Public Health Service Grant 5R01-AI09116, National Science Foundation Grant PCM78-18506 and the National Foundation of the March of Dimes 1-556.

References

1. Najjar, V. A. & Nishioka, K., 1970. Nature, 228: 672–673.
2. Najjar, V. A., 1974. Advances in Enzymology, Ed. A. Meister, Vol. 41, John Wiley & Sons, Inc., New York, pp. 129–278.
3. Najjar, V. A., 1980. Macrophages and Lymphocytes, Part A, Eds. M. R. Escobar & H. Friedman, Plenum Press, New York, pp. 131–147.
4. Nishioka, K., Constantopoulos, A., Satoh, P. S., Mitchell, W. M. & Najjar, V. A., 1973. Biochim. Biophys. Acta 310: 217–229.
5. Nishioka, K., Satoh, P. S., Constantopoulos, A. & Najjar, V. A., 1973. Biochim. Biophys. Acta 310: 230–237.
6. Fidalgo, B. V. & Najjar, V. A., 1967. Biochemistry, 6: 3386–3392.
7. Fidalgo, B. V. & Najjar, V. A., 1967. Proc. Natl. Acad. Sci. U.S.A. 57: 957–964.
8. Najjar, V. A., Fidalgo, B. V. & Stitt, E., 1968. Biochemistry, 7: 2376–2379.
9. Najjar, V. A. & Constantopoulos, A., 1972. J. Reticuloendothel. Soc. 12: 197–215.
10. Najjar, V. A., 1975. J. Pediat. 87: 1121–1124.
11. Najjar, V. A., 1978. Exp. Cell Biol. 46: 114–126.
12. Inada, K., Nemoto, N., Nishijima, A., Wada, S., Hirata, M. & Yoshida, M., 1980. Phagocytosis: Its Physiology and Pathology, Eds. Y. Kokobun & N. Kobayashi, Vol. 1, University Park Press, Baltimore, p. 157.
13. Fridkin, M., Stabinsky, Y., Zakuth, V. & Spirer, Z., 1977. Biochim. Biophys. Acta 496: 203–211.
14. Spirer, Z., Zakuth, V., Bogair, N. & Fridkin, M., 1977. Eur. J. Immunol. 7: 69–74.
15. Najjar, V. A. & Schmidt, J. J., 1980. Lymphokine Reports, Ed. E. Pick, Vol. 1, Academic Press Inc., New York, pp. 157–159.
16. Kabat, E. A., Wu, T. T. & Bilofsky, H., 1979. U.S. Dept. of Health, Education and Welfare, Public Health Service, National Institutes of Health #80-2008.
17. Babior, B. M., 1978. N. Eng. J. Med. 298: 659–668.
18. Sbarra, A. J., Selvaraj, R. J., Paul, B. B., Poskitt, P. K. F., Zgliezynski, J. M., Mitchell, G. W., Jr. & Louis, F., 1976. Int. Rev. Exp. Pathol. 162: 249–271.
19. Spirer, Z., Zakuth, V., Golander, A., Bogair, N. & Fridkin, M., 1975. J. Clin. Invest. 55: 198–200.
20. Martinez, J., Winternitz, F. & Vindel, J., 1977. Eur. J. Med. Chem.-Chimica Therapeutica, 12: 511–516.
21. Tzehoval, E., Segal, S., Stabinsky, Y., Fridkin, M., Spirer, Z. & Feldman, M., 1978. Proc. Natl. Acad. Sci. U.S.A. 75: 3400–3404.
22. Florentin, I., Bruley-Rosset, M., Imbach, J. L., Winternitz, F. & Mathé, G., 1978. Cancer Immunol. Immunother. 5: 211–216.
23. Nishioka, K., 1979. Br. J. Cancer, 39: 342–345.

24. McBride, J. K. & Stuart, A., 1977. The Macrophage and Cancer, Symposium of the European Reticuloendothelial Society, Ed. K. James, W. McBride & A. Stuart, University of Edinburgh, Edinburgh.

25. Klebanoff, S. J. & Clark, R. A., 1978. The Neutrophils: Functions and Clinical Disorders, Ed. S. J. Klebanoff & R. A. Clark, North-Holland Publishing Co., New York.

26. Catane, R., Schlanger, S., Gottlieb, P., Halpern, J., Treves, A. J., Fuks, Z. & Fridkin, M., 1981. Abstract.

27. Najjar, V. A., Chaudhuri, M. K., Konopinska, D., Beck, B. D., Layne, P. P. & Linehan, L., 1980. Augmenting Agents of Cancer Therapy, Ed. E. M. Hersh, M. A. Chirigos & M. Mastrangelo, Raven Press, New York. p. 459.

28. Stabinsky, Y., Fridkin, M., Zakuth, V. & Spirer, Z., 1978. Int. J. Peptide Protein Res. 12: 130–138.

29. Najjar, V. A., 1963. Physiol. Rev. 43: 243–262.

30. Baldwin, D., Jr., Terris, S. & Steiner, D. F., 1980. Journal of Biological Chemistry, 255: 4028–4034.

31. Najjar, V. A., 1951. Fed. Proc. (1951): 10: 227–228.

32. Zamecnik, P., Lipmann, F., 1947. J. Exp. Med. 85: 395–403.

33. Thomaidis, T. S., Fidalgo, B. V., Harshman, S. & Najjar, V. A., 1967. Biochemistry, 6: 3369–3377.

34. Oroszlan, S., Henderson, L. E., Stephenson, J. R., Copeland, T. D., Long, C. W., Ihle, J. N. & Gilden, R. V., 1978. Proc. Natl. Acad. Sci. U.S.A. 75: 1404–1408.

35. Luftig, R. B., Yoshinaka, Y. & Oroszlan, S., 1977. J. Cell Biol. 25: 399a.

36. Suk, W. A. & Long, C. W., 1979. Am. Soc. Microbiol. 257: 105.

37. Gethings, M. J., Bye, J., Skehel, J. & Waterfield, M., 1980. Nature, 287: 301–306.

38. Blundell, T. L. & Humbel, R. E., 1980. Nature 287: 781–787.

39. Malaisse-Lagae, R., Carpentier, J. L., Patel, Y. C., Malaisse, W. J. & Orci, L., 1977. Experienta, 33: 915–917.

40. Gates, R. J. & Lazarus, N., 1977. Hormone Res. 8: 189–202.

41. Konopinska, D., Nawrocka, E., Siemion, I. Z., Szymaniec, S. & Slopek, S., 1979. Arch. Immunol. Therap. Exper. 27: 151–157.

42. Hisatsune, K., Kobayashi, K., Nozaki, S. & Muramatsu, I., 1978. Microbiol. Immunol. 22: 581–584.

43. Veretennikova, N. I., Indulen, Yu. I., Nikiforovich, G. V., Papsuyevich, O. S. & Chipens, G. I., 1978. 11th Int. S. Chem. Nat. Vol. 1, p. 263–266.

44. Veretennikova, N. I., Chipens, G. I., Nikiforovich, G. V. & Betinsh, Ya. R., 1980. Int. J. Peptide Protein Res. in press.

Revision received February 12, 1981.

12

Antitumor effect of tuftsin

Kenji Nishioka,* George F. Babcock, Joseph H. Phillips,* and R. Dirk Noyes
*Dept. of General Surgery/Surgical Research Laboratory, The University of Texas System Cancer Center,
M. D. Anderson Hospital and Tumor Institute*
* *The University of Texas Health Science Center, Graduate School of Biomedical Sciences, Houston, Texas
77030, U.S.A.*

Summary

Tuftsin, a physiological tetrapeptide derived from the Fc region of leukophilic IgG possesses a variety of immunopotentiating properties including the ability to act as an immunotherapeutic agent against the experimental tumors, L1210 leukemia and Cloudman S-91 melanoma. Although the mechanism of action of tuftsin *in vivo* is not known, several types of leukocytes have been shown to become cytotoxic effector cells following activation with tuftsin. These cells presently include macrophages, natural killer cells, and granulocytes. The possibility that tuftsin can also activate other types of effector cells have not been ruled out. We feel this small peptide has a high potential (largely unrecognized) as an antitumor immunopotentiating agent. It is naturally occurring in man and appears to be relatively non-toxic. Its exact sequence (Thr-lys-Pro-Arg) is known and it can be chemically synthesized. Methods are also available to monitor the levels of tuftsin in body fluids. These properties along with its ability to control infectious disease make this agent one of the more promising immunopotentiators.

Introduction

Tuftsin, the physiological tetrapeptide (Thr-Lys-Pro-Arg) derived from the Fc portion of leukophilic immunoglobulin G (IgG) has been shown to stimulate the phagocytic activity of macrophages as well as neutrophils (1–3). Tuftsin was also shown to enhance the migration of granulocytes and peripheral blood mononuclear cells (1, 4, 5). It is well known that macrophages, monocytes and neutrophils play important roles as antineoplastic effector cells in animal and human hosts (6, 7). It is believed that the association of leukophilic IgG with Fc receptors induces a membrane protease to release tuftsin from the Fc portion of the immunoglobulin (1, 3). Tuftsin in turn can then bind to specific tuftsin receptors (8–10). Thus, leukocytes carrying Fc receptors could be the cells on which tuftsin acts. In this regard, natural killer (NK) cells are good candidates for tuftsin action as they are known to be antineoplastic effector cells and to possess low density Fc receptors (11). In addition, tuftsin stimulates many of the same biological activities as other immunotherapeutic agents. Examples of this include the levamisole-induced enhancement of granulocyte migration and the phagocytic activity of mononuclear phagocytes (12), and the stimulation of phagocytosis, pinocytosis and migration of BCG-activated macrophages (13, 14).

In view of this, it appeared warranted to examine the *in vivo* and *in vitro* antineoplastic properties of tuftsin. We thus began studies to determine if tuftsin has antitumor activity *in vivo* and if it could enhance the cytotoxic activity of the various leukocytes populations *in vitro*.

Antitumor effect *in vivo*

Our preliminary *in vivo* studies were initiated in 1976 using L1210 leukemia as a model. Tissue culture adapted leukemia cells (10^3–10^5 cells) were

Molecular and Cellular Biochemistry 41, 13–18 (1981).   0300-8177/81/0041-0013/$01.20.

14

injected intraperitoneally (i.p.) into syngeneic DBA/2 mice. In addition, experimental mice also received simultaneously tuftsin (0.2–20 μg). Tuftsin-treated mice showed significantly longer survival times than the untreated controls (15). Antitumor activity of tuftsin was also examined using Cloudman S-91 melanoma system. Experimental mice received 0.1 ml of cell suspension containing 2.5×10^5 tissue-cultured melanoma cells plus 10 μg tuftsin subcutaneously (s.c.) in the hind leg. Control mice were injected only with melanoma cells. The growth of melanoma was followed for 48 days by measuring the sizes of the palpable tumors. The melanomas in the experimental group grew significantly slower than those in the untreated group. In order to assess this effect more quantitatively, we have developed a melanoma colony assay in DBA/2 mice (16). In this system, tissue-cultured Cloudman S-91 melanoma cells were injected intravenously (i.v.) into the tail vein of the mouse. After an

appropriate period of time, pneumonectomy was performed and pigmented melanoma colonies in the lungs were then counted. In order to obtain numbers of colonies which can be counted at ease, 1×10^5 cells were injected and pneumonectomy performed approximately 5 weeks after the inoculation of tumor cells. In addition, profiles of two separate biochemical tumor markers, polyamines (16) and tyrosinase (17) were studied. A single dose (i.p.) of tuftsin did not show any therapeutic effect thus multiple dose regimens were deviced (18). Although we failed to observe a dose related response, tuftsin-treated mice showed significantly fewer colonies in the lungs as indicated in Table 1. It can be seen that 3 injections weekly for the duration of the experiment significantly reduced the number of melanoma colonies in the lungs of the treated mice over a wide range of dosages. In addition, tuftsin treatment started as late as 2 weeks after tumor inoculation still reduced the number of

Table 1. The effect of tuftsin on the development of Cloudman S-91 melanoma colonies in the lungs of DBA/2 mice[a]

No. of animals	Dosage of tuftsin (μg)	Frequency of administration	Mean no. of tumor colonies per mouse	P<[b]
		Experiment 1		
4	None	–	14	–
5	500	3×[c]	6	Ns[e] (0.1)
5	200	3× weekly[d]	0.05	0.02
		Experiment 2		
5	None	–	41	–
7	1	3× weekly[f]	13	0.005
7	5	3× weekly[f]	24	0.05
		Experiment 3		
5	None	–	385	–
5	50	3× weekly[d]	21	0.002
5	100	3× weekly[d]	74	0.005
		Experiment 4		
8	None	–	45	–
8	10	3× weekly[d]	5	0.0005
8	100	3× weekly[d]	18	0.005
8	1 000	3× weekly[d]	3	0.0005

[a] Mice received tumor i.v. on day 0, tuftsin treatment started on day 1. Mice were sacrificed and the tumors counted on day 31.
[b] Compared with Student's t-test.
[c] Given 3 times on days 1, 3 and 5.
[d] Given 3 times weekly until termination of experiment.
[e] Not significant.
[f] Treatment started 14 days after tumor implantation and the tumors counted on day 31.

colonies appearing in the lungs (Experiment 2). Serum tyrosinase appeared to be an effective marker to monitor the effect of tuftsin in this system. Recently, Bruly-Rosset *et al.* (19) have reported a preliminary study which indicates that treatment of aged mice with tuftsin significantly reduces the incidence of spontaneous tumorigenesis.

Antitumor effect *in vitro*

In order to elucidate the above mentioned *in vivo* antitumor effect of tuftsin, various *in vitro* experiments have been undertaken. To examine the possibility that tuftsin was directly cytotoxic to tumor cells, the effect of tuftsin on the incorporation of ^{3}H-thymidine by cultured L1210 leukemia cells was examined. Since the presence of tuftsin did not affect the incorporation of labeled thymidine (15), the antitumor activity of tuftsin is most likely to be exerted through immunological mechanisms at the effector cell level. We then examined the effect of tuftsin on the cytotoxicity of peritoneal macrophages from DBA/2 mice against L1210 cells. Monolayered peritoneal macrophages were activated for 18 h in microtiter plates with 1 or 10 μg/ml tuftsin. The macrophages were then washed and ^{3}H-proline labeled target (L1210 leukemia) cells were introduced into the wells at an effector: target cell ratio of 50:1. After 48 to 72 h incubation, the percent cytotoxicity was determined. The tuftsin-treated macrophages displayed significantly more cytotoxicity than the untreated macrophages (15). There are, however, other effector cells besides macrophages in peritoneal exudates (especially T cells) which could also be responsible for some of the effect observed *in vivo*.

Since alveolar macrophages probably represent the major class of effector cells in the lungs, similar experiments were performed to determine the effect of tuftsin on these cells. Alveolar macrophages were collected by gentle endobronchial lavage to prevent the introduction of blood into the cultures. The macrophages were further enriched by adherence to plastic. These plastic adherent cells were greater than 95% non-specific esterase positive. The aveolar macrophages were then incubated for 1 h *in vitro* with tuftsin. The macrophages were then incubated with ^{51}Cr-labeled Cloudman S-91 melanoma cells

and a standard 18 h ^{51}Cr release assay was performed. The tuftsin-stimulated macrophages displayed significant cytotoxicity for the melanoma cells while the untreated macrophages were not cytotoxic. Values of 50% cytotoxicity were reached at a ratio of 100:1 (effector vs target cells), which is particularly high for an 18 h assay using macrophages as effector cells.

We have also been able to demonstrate activation of alveolar macrophages *in vivo*. Our preliminary results indicate that at least 3 to 4 days are required to activate alveolar macrophages to become cytotoxic effector cells after i.p. injection of tuftsin. From our *in vitro* experiments, we know that a short exposure of tuftsin is sufficient to activate the effetor cells, however, the rather long period of time required for *in vivo* activation may indicate a more complex activation mechanism. There are some possible explanations which could be used to account for the rather long time period required for the *in vivo* tuftsin activation of effector cells, but we currently have no strong evidence to support or refute any of them. It is possible that it takes several days for a critical concentration of tuftsin to accumulate at the proper site *in vivo*. This is unlikely, however, as a long delay has also been observed for peritoneal cells activated *in vivo* (15, 20). A second possibility is that tuftsin stimulates an increase in the numbers of activated effector cells through mitosis. We have some evidence that tuftsin is mitogenic for plastic adherent cells (21). Since the lung is a source of stem cells (22), it is also possible that tuftsin causes stem cells or immature monocytes to differentiate into effector cells. We are currently pursuing these lines of investigation.

The effect of tuftsin on granulocytes was also examined using ^{3}H-proline labeled target cells (23). Human granulocytes from blood were purified, activated with tuftsin and tested against established human malignant melanoma cell line and a normal human embryonic lung cell line WI-38. Granulocytes demonstrated rather weak but significant cytotoxicity against human melanoma cell lines, but failed to show significant cytotoxicity against the normal cell line WI-38. Although untreated peritoneal macrophages always showed substantial natural cytotoxicity against tumor cells (15), granulocytes did not exhibit significant cytotoxicity without tuftsin. Microscopic examination of the microtiter wells showed no phagocytosis of tumor

16

Table 2. In vitro effects of tuftsin on murine splenic natural cell-mediated cytotoxicity[a]

Strain	Effectors	% Cytotoxicity ± S.E. Tuftsin (μg/ml)									
		0	.01	.1	1	10	50	100	200	500	1 000
CBA/J	Unfractionated	34 ± 1.0	32 ± 1.1	35 ± 1.0	46 ± 2.6	52 ± 1.5	57 ± 3.0	67 ± 1.0	62 ± 1.2	47 ± 2.0	42 ± 1.5
	Plastic nonadherent	25 ± 1.7	23 ± 1.1	27 ± 2.0	37 ± 1.5	57 ± 1.1	57 ± 2.0	62 ± 2.3	54 ± 1.5	36 ± 2.0	25 ± 2.5
	Nylon wool nonadherent	36 ± 1.0	32 ± 1.0	41 ± 2.0	50 ± 2.0	52 ± 3.0	62 ± 1.7	71 ± 2.0	72 ± 4.0	47 ± 2.0	29 ± 1.0
	Nylon wool nonadherent T-cells depleted	31 ± 2.0	25 ± 8.0	28 ± 2.0	52 ± 1.5	50 ± 3.0	56 ± 4.3	63 ± 2.5	63 ± 2.0	39 ± 2.5	26 ± 3.0
C57B1/10	Unfractioned	16 ± 1.5	14 ± 1.5	16 ± 2.0	29 ± 3.4	32 ± 3.2	34 ± 3.0	39 ± 3.0	37 ± 2.5	34 ± 1.0	28 ± 1.0
	Plastic nonadherent	12 ± 1.0	9 ± 1.5	10 ± 1.0	20 ± 1.0	27 ± 1.0	27 ± 1.0	32 ± 1.0	29 ± 1.0	22 ± 1.5	18 ± 1.3
	Nylon wool nonadherent	14 ± 1.0	13 ± 1.3	12 ± 1.5	22 ± 1.5	25 ± 1.0	29 ± 1.5	34 ± 2.0	32 ± 2.0	21 ± 1.2	20 ± 1.0
	Nylon wool nonadherent T-cell depleted	11 ± 1.0	11 ± 2.0	11 ± 1.5	23 ± 2.0	27 ± 2.0	28 ± 3.0	33 ± 2.5	35 ± 2.5	26 ± 2.0	22 ± 1.0
DBA/2	Unfractionated	14 ± 1.0	11 ± 2.0	12 ± 1.0	28 ± 3.0	32 ± 3.0	35 ± 3.0	38 ± 2.5	35 ± 2.3	30 ± 1.0	25 ± 1.5
	Plastic nonadherent	14 ± 1.0	13 ± 1.0	13 ± 1.1	27 ± 1.0	34 ± 1.0	33 ± 1.0	38 ± 3.0	37 ± 2.0	27 ± 1.0	20 ± 1.0
	Nylon wool nonadherent	17 ± 1.5	18 ± 1.0	18 ± 1.0	32 ± 1.7	39 ± 1.5	39 ± 2.0	45 ± 2.0	42 ± 1.5	31 ± 1.5	25 ± 1.5
	Nylon wool nonadherent T-cell depleted	13 ± 1.5	13 ± 1.2	11 ± 1.5	20 ± 1.0	28 ± 1.5	33 ± 1.1	38 ± 2.0	37 ± 2.5	24 ± 1.0	17 ± 1.5

[a] Effector cells were incubated for 1 h at 37 °C with various concentrations of tuftsin before the addition of target cell. NK cell cytotoxicity was assessed in an 18 h ^{51}Cr release assay against the T-cell lymphoma, Yac-1 at an effector to target cell ratio of 50:1. The results are expressed as the mean percent cytotoxicity ± standard error of the means. Significant enhancement ($p < 0.005$) of NK cell cytotoxicity over untreated controls was observed in all effector populations at tuftsin concentrations of 1–500 μg/ml.

cells by the granulocytes indicating that the granulocyte cytotoxicity was not expressed by enhanced phagocytosis.

We have also initiated studies to investigate the effects of tuftsin on NK-cell mediated cytotoxicity (21). The physiological significance of NK cells is not precisely known. However, evidence has now accumulated implicating NK cells as an effector mechanism in the immunoresistance to neoplastic growth (11). In view of this, we have undertaken a series of studies to investigate the *in vitro* effects of tuftsin on murine and human NK cell cytotoxicity. Table 2 summarizes the effect of tuftsin on murine splenic NK cell cytotoxicity. The results of these studies clearly indicate that the *in vitro* treatment of murine splenic effector cells with tuftsin induced a significant enhancement of NK cell cytotoxicity against the allogeneic A strain T-cell lymphoma Yac-1 cell line. The magnitude of stimulation of NK cell cytotoxicity was dependent upon the concentration of tuftsin employed. Consistant stimulation of NK cell cytotoxicity in three strains of mice was observed at tuftsin concentrations of 1–500 μg/ml. The dose-dependent stimulation of NK cell cytotoxicity with tuftsin appears to indicate a receptor-

mediated phenomenon with maximum enhancement at 50–100 μg tuftsin per ml. Removal of macrophages, monocytes, B-cells and T-cells from effector populations had no effect on the tuftsin-induced enhancement of cytotoxicity, thus clearly demonstrating a direct effect on tuftsin on NK cell enriched populations.

Recently, we have obtained data to indicate that *in vitro* treatment of normal human peripheral blood effector cells with tuftsin also induces a pronounced enhancement of natural cell-mediated cytotoxicity against human melanoma cell line 39-5 as mediated by three separate effector populations (Table 3).

Other relevant effects and prospect

Florentin *et al.* (20) showed that i.v. injection of tuftsin into mice rendered the peritoneal macrophages highly cytostatic against L1210 leukemia cells. They also observed enhancement of antibody-dependent cell-mediated cytotoxicity of spleen cells against chicken red blood cells and potentiation of the antibody response to a thymus-dependent anti-

Table 3. In vitro effects of tuftsin on human peripheral blood natural cell-mediated cytotoxicity[a]

Effectors	% Cytotoxicity ± S.E. Tuftsin (µg/ml)			
	0	10	50	100
Mononuclear cells	8.0 ± 1.0	20 ± 2.0	32 ± 3.0	24 ± 1.3
Plastic nonadherent	10.0 ± 1.3	23 ± 2.0	31 ± 3.0	25 ± 1.5
Macrophages/monocytes	1.4 ± 2.0	13 ± 2.0	20 ± 1.5	16 ± 2.1
Polymorphonuclear leukocytes	2.7 ± 1.0	15 ± 2.6	26 ± 3.0	18 ± 2.1

[a] Effector cells were incubated for 1 h at 37 °C with various concentrations of tuftsin before the addition of target cells. NK cell cytotoxicity was assessed in an 18 h ^{51}Cr release assay against the human melanoma HM 39.5 at an effector to target ratio of 50:1. The results are expressed as the mean percent cytotoxicity ± the standard error of the means. Significant enhancement ($p < 0.005$) of NK cell cytotoxicity over untreated controls was observed in all four effector populations at tuftsin concentrations of 10–100 µg/ml.

gen by tuftsin. Tzehoval *et al.* (24) demonstrated augmentation of the antigen-specific, macrophage-dependent education of T-lymphocytes. We have observed the morphological alteration or spreading of mouse peritoneal macrophages, and increased numbers of leukocytes in the peritoneal exudate following i.p. injection of tuftsin (15), and *in vitro* migration enhancement of human mononuclear cells by tuftsin (5). Since tuftsin induced an increase in leukocyte count in peritonium suggesting that it may possess chemotactic activity, we have carried out systematic study of chemotaxis using blind well chambers (25). Our study indicated that tuftsin clearly possessed chemotactic activity for human blood mononuclear leukocytes and granulocytes, whereas the peptide Kenstin (26), which has the same amino acid composition as tuftsin, but a different sequence (Thr-Pro-Arg-Lys) exhibited no such effect. Due to the known antineoplastic effect of interferon (27), the effect of tuftsin on interferon induction was also examined using the vesicular stomatitis virus plaque reduction assay in mouse cells. No effect was observed. In addition, the effect of tuftsin on the release of lysosomal enzyme (28) from human leukocytes was examined as a possible mechanism for antitumor activity of tuftsin. Again, no effect was seen. The various effects of tuftsin on the immune system may contribute to the *in vivo* antitumor activity of tuftsin. However, elucidation of mechanism of antitumor effect of tuftsin requires further investigation.

Frequent presence of infections among patients with tumors, the availability of a radioimmunoassay for tuftsin which permits measurements of serum tuftsin levels at any given time, and the expected low toxicity of tuftsin (18) make tuftsin a very attractive agent for cancer immunotherapy.

Acknowledgements

This work was partly supported by grants CA 27330 and CA 16672 awarded by NCI, DHEW. The authors wish to acknowledge and thank Dr. James Chan for the interferon assays, Dr. Marvin Romsdahl for his encouragement, Mark Rosen for experiments with alveolar macrophages and Laura Lerma for her excellent secretarial work in preparing the manuscript.

References

1. Nishioka, K., Constantopoulos, A., Satoh, P. S. & Najjar, V. A., 1972. Biochem. Biophys. Res. Commun. 47: 172–179.
2. Constantopoulos, A. & Najjar, V. S., 1972. Cytobios 6: 97–100.
3. Nishioka, K., Constantopoulos, A., Satoh, P. S., Mitchell, W. M. & Najjar, V. A., 1973. Biochim. Biophys. Acta 310: 217–229.
4. Nishioka, K., Satoh, P. S., Constantopoulos, A. & Najjar, V. A., 1973. Biochim. Biophys. Acta 310: 230–237.
5. Nishioka, K., 1978. Gann 69: 569–572.

18

6. James, K., McBride, B. & Stuart, A., 1977. The Macrophage and Cancer. University of Edinburgh, Edinburgh.

7. Klebanoff, S. J. & Clark, R. A., 1978. The Neutrophil: Function and Clinical Disorders. North-Holland Publishing Co., New York.

8. Stabinsky, Y., Gottlieb, P., Zakuth, V., Spirer, Z. & Fridkin, M., 1978. Biochem. Biophys. Res. Commun. 83: 599–606.

9. Nair, R. M. G., Ponce, B. & Fudenberg, H. H., 1978. Immunochem. 15: 901–907.

10. Bar-shavit, Z., Stabinsky, Y., Fridkin, M. & Goldman, R., 1979. J. Cell. Physiol. 100: 55–62.

11. Kiessling, R. & Wigzell, H., 1979. Immunol. Rev. 44: 1–5–208.

12. Anderson, R., Glover, A., Koornhof, H. J. & Rabson, A. R., 1976. J. Immunol. 117: 428–432.

13. Scheetz, M. E. II, Thomas, L. J., Allemenos, D. K. & Schinitsky, M. R., 1976. Immunol. Commun. 5: 189–203.

14. Poplack, D. G., Sher, N. A., Chaparas, S. D. & Blaese, R. M., 1976. Cancer Res. 36: 1233–1237.

15. Nishioka, K., 1979. Br. J. Cancer 39: 342–345.

16. Takami, H. & Nishioka, K., 1980. Br. J. Cancer. In press.

17. Nishioka, K., Romsdahl, M. M., Fritsche, H. A. Jr. & McMurtrey, M. J., 1977. Int. J. Cancer 20: 689-693.

18. Noyes, R. D., Babcock, G. F. & Nishioka, K., 1980. Proc. Am. Assoc. Cancer Res. 21: 261.

19. Bruley-Rosset, M., Hercend, T., Florentin, I. & Mathé, G., 1980. Proc. Am. Assoc. Cancer Res. 21: 250.

20. Florentin, I., Bruley-Rosset, M., Kiger, N., Imback, J. L., Winternitz, F. & Mathé, G., 1978. Cancer Immunol. Immunother. 5: 211–216.

21. Babcock, G. F., Phillips, J. H. & Nishioka, K., 1980. 4th Int. Congress Immunol. In press.

22. Reppun, T., Lin, S. & Kuhn, C., 1979. J. Reticuloendothel. Soc. 25: 379–386.

23. Nishioka, K., 1979. Gann 70: 845–846.

24. Tzehoval, E., Segal, S., Stabinsky, Y., Fridkin, M., Spirer, Z. & Feldman, M., 1978. Proc. Natl. Acad. Sci. USA 75: 3400–3404.

25. Boumsell, L. & Meltzer, M. S., 1976. J. Immunol. 115: 1746–1748.

26. Kent, H. A., 1974. Biol. Reprod. 12: 504–507.

27. Friedman, R. M., 1978. J. Natl. Cancer Inst. 60: 1191–1194.

28. Becker, E. L. & Showell, H. J., 1974. J. Immunol. 112: 2055–2062.

Received July 24, 1980.

Nα-formyl-norleucyl-leucyl-phenylalanyl chloromethylketone
A possible covalent agonist of the Nα-formyl-methionyl-peptide receptor

Henry J. Showell, William M. Mackin and Elmer L. Becker
Dept. of Pathology, University of Connecticut Health Center, Farmington, Conn. 06032, U.S.A.

Nakesa Muthukumarasway, Alan R. Day and Richard Freer
Dept. of Pharmacology, Medical College of Virginia, Richmond, VA 23298, U.S.A.

Summary

Nα-formyl-norleucyl-leucyl-phenylalanine-chloromethyl ketone is chemotactic for, and induces lysosomal enzyme release from rabbit peritoneal neutrophils over essentially the same range of concentrations as does the free acid form of the same peptide (Nα-formyl-norleucyl-leucyl-phenylalanine-OH). The chloromethyl ketone derivative does however differ from the free acid in respect to its ability to interact with the neutrophil and cause deactivation or desensitization to cytochalasin B. Neutrophils preincubated in the cold with the chloromethyl ketone followed by washing have cytochalasin B sensitivity conferred upon them, as measured by the release of lysosomal enzymes. The degree of release induced by this pre-treatment appears to be related to the initial responsiveness of the cells. This is in contrast to the free acid where no cytochalasin B sensitivity in conferred under any circumstances. Thus, the chloromethyl ketone, unlike the free acid, appears to irreversibly activate the cell. Desensitization to the late addition of cytochalasin B is also significantly retarded when the chloromethyl ketone derivative is compared to the free acid form of the peptide. These studies suggest that the chloromethyl ketone derivative of the peptide may covalently interact with the neutrophil receptor.

Introduction

Synthetic formylated oligopeptides interact with a specific receptor on neutrophils (1). A ligand that could covalently interact with the binding site of the neutrophil receptor i.e. affinity label it, would be useful in the identification and purification of this receptor. A similar procedure has been successfully used to identify and purify the acetylcholine receptor (2). Recently, Niedel has reported preliminary evidence that three different affinity-labeled derivatives of formyl-norleucyl-leucyl-phenylalanyl-norleucyl-tyrosyl-lysine, (f Nle-Phe-Nle-Tyr-Lys) bind covalently to human neutrophils, presumably at the formyl peptide receptor (3).

The observation that formylmethionyl-leucyl-descarboxy phenylalanine, (f Met-Leu-Pea), an analog of formylmethionyl-leucyl-phenylalanine, (f Met-Leu-Phe), incapable of either ionic or hydrogen bonding interactions at the C-terminus, was only poorly active (4) suggested an important role for the carboxyl portion of the ligand in receptor interaction. This, coupled with the extremely effective use of amino acid and peptide chloromethylketones in generating active-site directed enzyme antagonists (5), suggested that synthesis of a tripeptide N-formyl-norleucyl-leucyl-phenylalanyl-chloromethylketone, (f Nle-Leu-Phe-COCH$_2$Cl) might result in an active-site directed receptor antagonist. In this paper, we report the synthesis of f Nle-Leu-Phe-COCH$_2$Cl and compare the binding and biological activity of the chloromethyl ketone and the homologous, formyl-norleucyl-leucyl-phenylalanine, free acid (f Nle-Leu-Phe). In addition we report a heretofore

Molecular and Cellular Biochemistry 41, 19–25 (1981).   0300–8177/81/0041–0019/$01.40.

unrecognized relationship between the sensitivity of neutrophils to a formylpeptide and the extent to which it can become deactivated by the same peptide, as measured by cytochalasin B dependent lysosomal enzyme secretion.

Methods

Starting materials for the synthesis of the peptides were formyl-norleucyl-leucine, (f Nle-Leu-OH), prepared by rapid mixed anhydride coupling of t-butoxycarbonyl-norleucine (t-Boc-Nle) and leucine benzyl ester (leu-OBzl) (6). The resulting dipeptide was de-protected with trifluoroacetic acid and the formyl group introduced via a mixed anhydride coupling using isovaleryl chloride (Day et al. unpublished data). The benzyl group was removed by anhydrous HF (7). Carbobenzoxy phenylalanyl-chloromethylketone, (CBZ-Phe-COCH$_2$Cl) was purchased from Sigma Chemicals and the carbobenzoxy group removed by HBr/acetic acid. Each intermediate was completely characterized and judged homogenous by thin layer chromatography (tlc) in three different solvent systems.

The final product, f Nle-Leu-Phe-COCH$_2$Cl was prepared by coupling the f Nle-Leu-OH and H-Phe-OCH$_2$Cl again using the method of Tilak (6). The compound was obtained in 91% yield by crystallization from ethyl acetate. The chloromethyl ketone compound was also judged homogeneous in 3 tlc systems using Merck Silica gel F-254 glass plates, R$_f$ Sys A = 0.77 (Sys. A = n butanol: acetic acid: H$_2$O 4:1:1) R$_f$ Sys B = 0.73 (Sys B = benzene: H$_2$O: acetic acid, 9:1:9) and R$_f$ Sys C = 0.90 (Sys C = chloroform: methanol: acetic acid: H$_2$O, 60:30:4:1). The compound exhibited a sharp melting point at 189° and the expected 1:1 ratio of Nle and Leu on amino acid analysis. Elemental analysis showed C, H, N, Cl expected (found), C, 61.11(61.09): H,

7.59(7.68): N, 9.29(9.14) and Cl, 7.84(8.03). $[\alpha]_D^{20} = $ -93.8(C = 1, DMF) [^{3}H]FMLP binding studies were carried out on rabbit neutrophils, as previously reported, using f Met-Leu-[^{3}H]Phe (New England Nuclear Co., Boston MA) or f Nle-Leu-[^{3}H]Phe (1, 8). Cells were preincubated with the f Nle-Leu-Phe-COCH$_2$Cl or f Nle-Leu-Phe for 30 min at 0 °C, washed three times with ice cold buffer, and then assayed for [^{3}H]FMLP binding.

Chemotaxis and lysosomal enzyme release were performed as described (9). Although, only the results of lysozyme release are shown, the results with β glucuronidase were parallel. Cell viability was monitored by assaying for lactate dehydrogenase activity. In all cases, enzyme leakage did not exceed 5% of the total, indicating good cell viability. Hanks balanced salt solutions with 10 mM HEPES (N-2-Hydroxyethyl peperazine-N'-2 ethanesulfonic acid) buffer at pH 7.2 were used throughout for these studies.

Results

Table 1 lists the ED50s (concentration giving 50% of maximal response) for chemotaxis and lysosomal enzyme secretion and the ID50s (concentration inhibiting 50% of binding) for inhibition of binding of f Nle-Leu-[^{3}H]Phe with both the free acid and the chloromethyl ketone derivative of f Nle-Leu-Phe. As can be seen, the two compounds differ only slightly in their ability to initiate neutrophil migration or to inhibit the binding of f Nle-Leu-[^{3}H]Phe, indicating that the chlormethyl ketone is almost as active as the free acid in interacting at the formylpeptide receptor and inducing biological activity.

We have shown that neutrophils preincubated at 37° with f Met-Leu-Phe at concentrations greater than 10^{-10} M become hyporesponsive (deactivated) to a second challenge of the same peptide (10).

Table 1.

	Chemotaxis		Lysozyme release		^{3}Hf Nle-Leu-Phe binding
	ED50(M)	n	ED50(M)	n	ID50(M)
f Nle-Leu-Phe-OH	5 × 10^{-10}	1	1.34 ± 0.45 × 10^{-9}	4	8 × 10^{-9}
f Nle-Leu-Phe-COCH$_2$Cl	1.15 ± 0.5 × 10^{-9}	2	1.52 ± 0.31 × 10^{-9}	4	2.5 × 10^{-8}

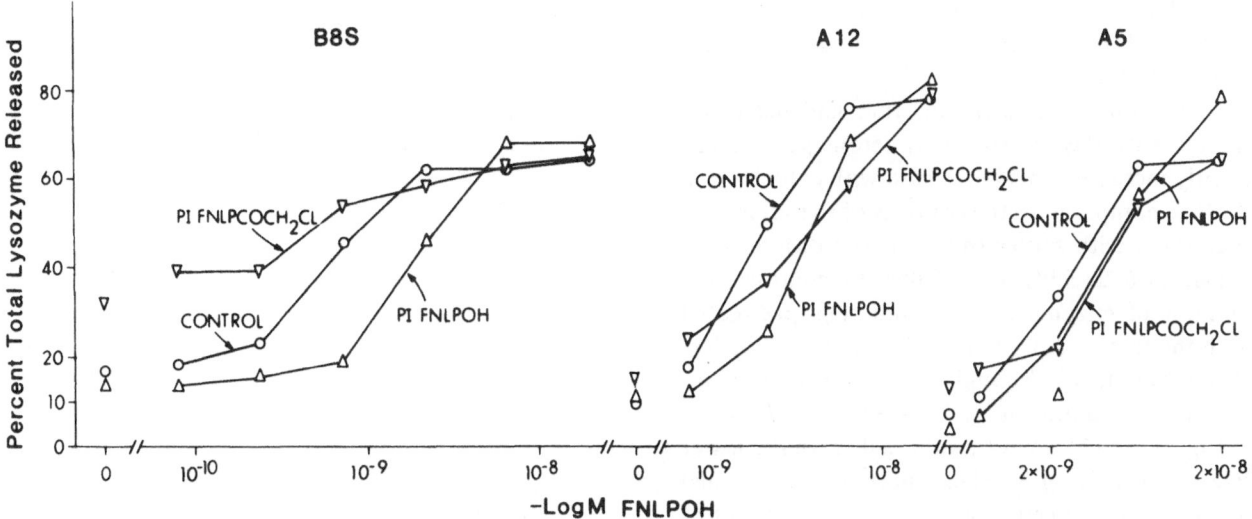

Fig. 1. Lysozyme release induced by serial dilutions of f Nle-Leu-Phe from three different rabbits, after preincubation (PI) at 0 °C for 30 min, with f Nle-Leu-Phe 10⁻⁶ M (△--△), f Nleu-Leu-Phe-COCH₂Cl 10⁻⁶ M (▼--▼) or control (no addition) (○--○) followed by washing.

When pretreated cells are compared to control cells for their ability to give peptide induced lysosomal enzyme secretion in the presence of cytochalasin B, the degree of deactivation is seen as a shift to the right in the dose-response curve. Neutrophils preincubated at 0 °C instead of 37 °C with saturating concentrations of peptide (10^{-6} M to 10^{-8} M) also become deactivated. As shown in Fig. 1, preincubating neutrophils at 0° for 30 min with 10^{-6} M f Nle-Leu-Phe shifts the dose response curves to the right for three different batches of neutrophils. To be especially noted, is that neutrophils pretreated with 10^{-6} M f Nle-Leu-Phe-COCH₂Cl then washed, released significantly higher amounts of enzyme on the addition of cytochalasin B alone than did controls pretreated with buffer. In contrast, neutrophils pretreated with f Nle-Leu-Phe gave no greater release of granule enzymes upon the addition of cytochalasin B alone than did the controls (Fig. 1). In addition, the untreated cells from rabbits A12 and A5 were significantly less sensitive responders to f Nle-Leu-Phe (ED50s, 1.8×10^{-9} M and 2.8×10^{-9} M), respectively than those from rabbit B8S (ED50, 5.8×10^{-10} M) and after pretreatment with f Met-Leu-Phe-COCH₂Cl, they gave much less release of lysozyme on the addition of cytochalasin B alone than did the cells from B8S (13.5 and 15.5 percent vs

32.5%). To see if the apparent relation between initial responsiveness and ability to give f Nle-Leu-Phe-COCH₂Cl induced release held, six more rabbits were investigated in the same way. The results obtained with the neutrophils of all nine rabbits are shown in Fig. 2 where the ED50s of the cells treated with f Nle-Leu-Phe plus cytochalasin B are plotted against the percent release of lysozyme induced by cytochalasin B alone from cells pre-

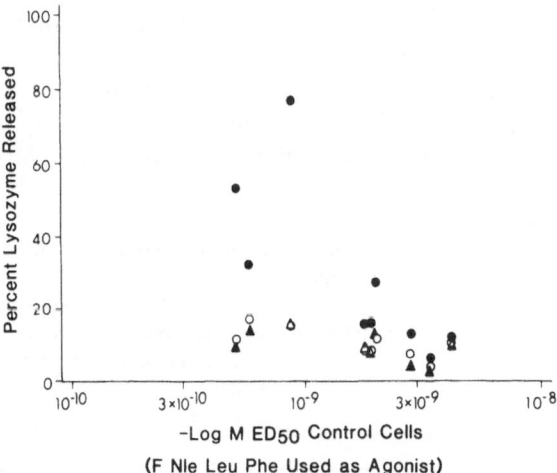

Fig. 2. Cytochalasin B induced lysozyme release from neutrophils of nine different rabbit preincubated at 0 °C for 30 min with f Nle-Leu-Phe 10⁻⁶ M (▲), f Nle-Leu-Phe-COCH₂Cl 10⁻⁶ (●) or control (no additions) (○) followed by washing.

treated at 0 °C for 30 min with buffer (control cells), 10^{-6} M f Nle-Leu-Phe, or 10^{-6} M f Nle-Leu-Phe COCH$_2$Cl.

Cells from all nine rabbits that had been pretreated with the chloromethyl ketone gave significantly increased enzyme release with cytochalasin B alone compared to the controls of the same cells pretreated with buffer (mean paired difference ± SEM, 17.6 ± 6.9% P > 0.05). In contrast, the amount of enzyme release from cells pretreated with the free acid, f Nle-Leu-Phe tended to be less than the controls, although the differences were not statistically significant (-1.1 ± 0.81%, p > 0.1).

Figure 2 also shows that the cells with the lower ED50 for f Nle-Leu-Phe, (more sensitive cells) tended to be the cells that gave a higher extent of enzyme release after pretreatment with the chloromethylketone compound and the addition of cytochalasin B alone. Although less evident, the same tendency seems to be present in the control cells and in cells pretreated with f Nle-Leu-Phe. The correlation co-efficients (r) between ED50 and percent enzyme release following cytochalasin B treatment for cells pretreated with f Nle-Leu-Phe-COCH$_2$Cl, f Nle-Leu-Phe or buffer (control) are -0.65, -0.64 and -0.59, respectively. In no case, are the correlation coefficients statistically significant, although they approach significance quite closely. Thus, although the correlation between initial sensitivity to peptide and ability to respond to cytochalasin B is probably real, more experiments are required to unequivocally demonstrate it.

However, as seen in Fig. 3, there is no doubt that neutrophils which respond more sensitively to f Nle-Leu-Phe (lower ED50) are also the neutro-

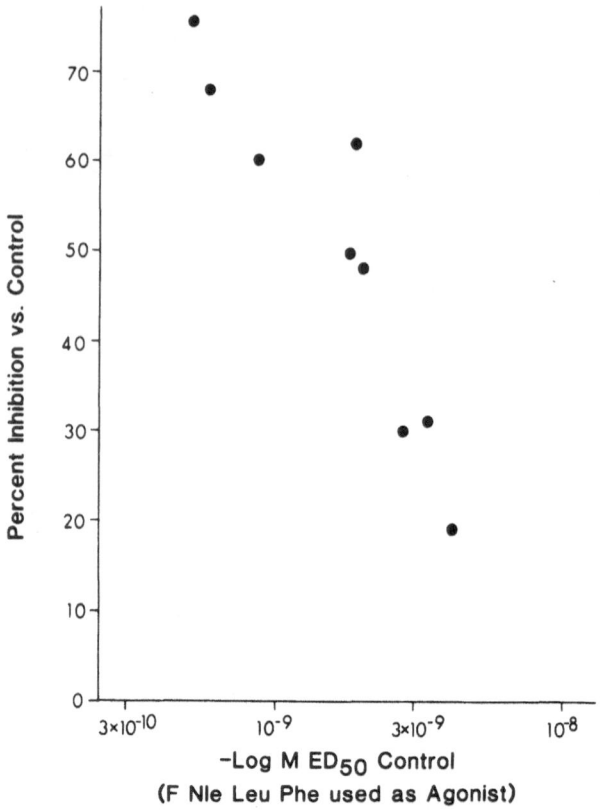

Fig. 3. Inhibition of lysozome release induced in neutrophils from nine different rabbits, after preincubation for 30 min at 0 °C with 10^{-6} M f Nle-Leu-Phe for a given population of cells is plotted against the ED$_{50}$ for f Nle-Leu-Phe obtained from control cells of the same batch.

phils which are deactivated to a greater extent by saturating concentrations (10^{-6} M) of f Nle-Leu-Phe (r = 0.95).

We next tested the extent of maximum specific binding of f Met-Leu-[^{3}H]Phe to neutrophils from

Table 2. Effect of incubation with chloromethyl ketone and homologuous free acid on loss of binding activity.

Rabbit #	Lysozyme release by f Nle-Leu-Phe ED50	Specific binding (CPM × 10³/10⁷ cells)				
		Control	f Nle-Leu-Phe (Inhibition)		f Nle-Leu-Phe-COCH$_2$Cl (Inhibition)	
			10^{-6} M	%	10^{-6} M	%
B5	4.2×10^{-9} M	11.7	7.87	(33)	5.95	(49)
A39	1.9×10^{-9} M	19.8	14.29	(28)	8.76	(56)
A42	3.4×10^{-9} M	28.0	14.87	(47)	12.26	(56)
SE42	5.1×10^{-10} M	11.1	6.6	(45)	4.98	(56)
455	8.7×10^{-10} M	14.6	9.11	(38)	4.89	(67)

five of the nine rabbits after pretreatment at 0 °C with either buffer, 10^{-6} M f Nle-Leu-Phe or 10^{-6} M f Nle-Leu-Phe-COCH$_2$Cl. Table 2 shows that pretreatment of neutrophils with f Nle-Leu-Phe at 0° for 30 min led to a loss of binding sites for f Met-Leu-[^{3}H]Phe. This down regulation at 0° confirms the findings of Sullivan and Zigmond (11). However, in every instance, pretreatment with the chloromethyl ketone caused a greater drop in specific binding, than that found with the free acid ($p < 0.01$). Table 2 however, reveals no obvious relation between the reduction in binding sites by f Nle-Leu-Phe and the sensitivity of response of the neutrophils (ED50) and by implication (see Fig. 3), the extent of deactivation.

Exposure of neutrophils to chemotactic factor followed 15 s to 4 min later by the addition of cytochalasin B leads to a diminished response when compared to the response obtained when cells are challenged simultaneously with chemotactic agent and cytochalasin B (10). If the interval is prolonged to 5 min and the neutrophils are in a buffer lacking Ca^{2+} there is a partial return of the responsiveness. Using the latter procedure with the neutrophils in

the absence of added Ca^{2+}, as described in (10), f Nle-Leu-Phe and f Nle-Leu-Phe-COCH$_2$Cl showed a distinct separation in the pattern of their responses (Fig. 4). The free acid f Nle-Leu-Phe behaved as already reported for f Met-Leu-Phe (10). The response obtained when the f Nle-Leu-Phe-COCH$_2$Cl preceded the addition of cytochalasin B, although uniformly diminished compared to that obtained from the simultaneous addition of the chloromethyl ketone and cytochalasin B, was nevertheless three to four times greater at the three time points tested than the response to the free acid under corresponding circumstances. Identical results were observed with the cells obtained from three other rabbits.

Discussion

In the above, we have considered two separate, although related aspects of the secretory response of neutrophils to synthetic formyl peptides. The first is the comparison of the responses of the chloromethyl ketone derivative, f Nle-Leu-Phe-

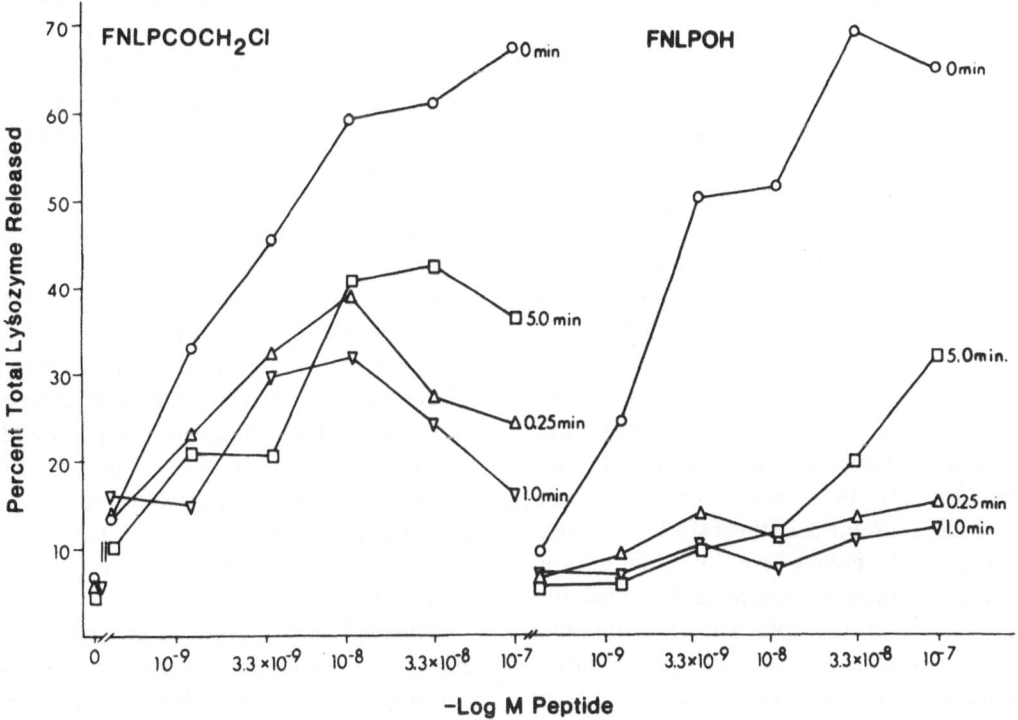

Fig. 4. Inhibition of lysozyme, release induced by late addition of cytochalasin B as seen with either f Nle-Leu-Phe or f Nle-Leu-Phe-COCH$_2$Cl. Cytochalasin B added at 0 min (same time as peptide) O–O, 0.25 min after peptide △–△, I min after peptide ▽–▽ and 5 min after peptide □–□.

COCH$_2$Cl and the free acid f Nle-Leu-Phe. The second aspect is the relationship of what one might call the intrinsic sensitivity of a given neutrophil population to the extent to which it can be deactivated. The two aspects will be treated separately in what follows.

Comparison of the behavior of the chloromethyl ketone and the free acid

Although we obviously have no direct evidence that the chloromethyl ketone binds covalently to the binding site of the formyl peptide receptor, certain of our results can be most easily explained on this basis. The strongest support for this idea is the finding that incubation of neutrophils with the chloromethyl ketone in the cold and then washing out the unbound peptide yields a cell that is apparently irreversibly 'turned on'; that is, these pretreated neutrophils in the absence of added chemotactic factor, on the mere addition of cytochalasin B will secrete granule enzymes to a distinctly greater degree than before treatment. Cells treated in the same manner with the free acid, f Nle-Leu-Phe, show no corresponding increase in secretory capacity even though the free acid is possibly slightly more active than the chloromethylketone in inducing locomotion, preventing the binding to the receptor of f Met-Leu-[^{3}H]Phe and as active in inducing granule enzyme secretion (Table 1). The irreversible turning on of the neutrophil by the chloromethyl ketone is analogous to the irreversibly stimulated discharge of secretory proteins found when the photoaffinity analogue of the peptide pancreatic hormone, cholecystokinin was reacted with the acinar cells of the pancreas (12).

The formyl peptides and C5a down regulate their neutrophil receptors i.e. they cause them to lose binding activity (13, 14). They also deactivate i.e. induce a loss of biologic responsiveness in the neutrophil (10, 15). Both down regulation and deactivation have complicated these studies, forcing us to pretreat in the cold where down regulation and deactivation though present is less evident. Nevertheless, even in the cold, the chloromethyl ketone causes a greater loss of neutrophil binding sites than does the free acid. This difference is also compatible with an irreversibly covalent binding of the chloromethyl ketone to the receptor. However, the relatively small differences suggest that the covalent binding, if it occurs under these circumstance, is slow and incomplete, possibly, in part, because of the lower temperature.

Neutrophils pretreated with the chloromethyl ketone are not as susceptible to desensitization to cytochalasin B as are those pretreated with the free acid Fig. 4. A difficulty with trying to explain the lessened ability of the chloromethyl ketone compared to the free acid to desensitize to cytochalasin B is that we have, at present, no well founded explanation of the basic phenomenon itself. Thus, any speculation we advance would have to be based on a further speculation as to the mechanism of desensitization to cytochalasin B. We shall mercifully spare the reader this piling of Ossa on Pelian. Nevertheless, the finding with regard to desensitization to cytochalasin B, like those of Table 1, tends to rule out the idea that the chloromethyl ketone is merely more generally active than the homologuous free acid and in this negative sense strengthens the explanation we have advanced for the ability of the chloromethyl ketone to irreversibly turn on the cell.

Thus, although we have no direct evidence that the chloromethyl ketone binds covalently to the formyl peptide binding site, the varied differences we have noted between the chloromethyl ketone and the free acid are compatible with this idea. They provide encouragement to synthesize more active derivatives of the same kind which could be iodinated or tritiated to serve as affinity labels for the receptor.

The relation of neutrophil responsiveness to effects of neutrophil manipulation

It is clear from Fig. 3 that under the condition of receptor saturation the extent of deactivation is directly related to the initial responsiveness of the cells, and from Table 2 that it is independent of the degree of down regulation. These findings are in accord with our hypothesis that homologuous deactivation is due not only to receptor down regulation but also to one or more postreceptor events (13).

One explanation of the relationship portrayed in Fig. 3 is that the sensitivity of response of a given cell population to the chemotactic factor depends upon the availability of a critical metabolite or metabolites or activity of a critical enzyme. The greater the availability or activity the greater the loss during the deactivation process and the greater the extent of the deactivation. An alternative explanation is that the more responsive the cell the greater the stimulation of the production of substances that decrease the ability of neutrophils to respond to further stimulation. These substances may be oxidative metabolites, as postulated by Nelson *et al.* (16) or granule contents as suggested by Gallin *et al.* (17). At present we are in no position to choose between these two types of explanation.

However, under the appropriate circumstances, not only is the degree of deactivation a function of the initial responsiveness of the cell but the results of Fig. 2 suggest that the ability of the cell to secrete granule enzymes in response to cytochalasin B alone also correlates with the initial responsiveness of the cell. If further experiments were to bear out this latter suggestion this would support the first type of explanation offered for the relation between responsiveness and deactivation. In addition, it would suggest that the secretion induced in rabbit peritoneal neutrophils (18) by cytochalasin B alone depends, in part, on a critical process or processes which are also responsible for secretion induced by chemotactic factor and cytochalasin B.

Acknowledgements

This work was supported by Grants AI 09648, NIH Contract DE-62494 and NIH S T32 HL07202.

References

1. Aswanikumar, S., Corcoran, B. A., Schiffmann, E., Day, A. R., Freer, R. J., Showell, H. J., Becker, E. L. & Pert, C. B., 1977. Biochem. Biophys. Res. Commun. 74: 810–817.

2. Nathanson, N. N. & Hall, Z. W., 1980. J. Biol. Chem. 255: 1698–1703.
3. Niedel, J., Davis, J. & Cuatrecasas, P., 1980. J. Biol. Chem. 255: 7063–7066.
4. Freer, R. J., Day, A. R., Radding, J. A., Schiffmann, E., Aswanikumar, S., Showell, H. J. & Becker, E. L., 1980. Biochem. 19: 2404–2410.
5. Shaw, E., 1980. Enzyme inhibitors as drugs. (Sandler, M., ed.), pp. 25–42. Univ. Park Press, Baltimore.
6. Tilak, M. A., 1970. Tetrahedron Lett. 11: 849–854.
7. Stewart, J. M. & Young, J. D., 1969. Solid-Phase Peptide Synthesis, (Freeman, W. H., ed.), San Francisco, CA.
8. Day, A. R., Radding, J. A., Freer, R. J., Showell, H. J., Becker, E. L., Schiffmann, E., Corcoran, B., Aswanikumar, S. & Pert, C. B., 1977. Febs. Lett. 77: 291–294.
9. Showell, H. J., Freer, R. J., Zigmond, S. H., Schiffmann, E., Aswanikumar, S., Corcoran, B. & Becker, E. L., 1976. J. Exp. Med. 143: 1154–1169.
10. Showell, H. J., Williams, D., Becker, E. L., Naccache, P. H. & Sha'afi, R. I., 1974. J. Ret. End. Soc. 25: 139–150.
11. Sullivan, S. J. & Zigmond, S. H., 1980. J. Cell. Biol. 85: 703–711.
12. Galardy, R. E. & Jamieson, J. D., 1977. Mol. Pharm. 13: 852–863.
13. Vitkauskas, G., Showell, H. J. & Becker, E. L., 1980. Mol. Immunol. 17: 171–180.
14. Chenoweth, D. E. & Hugli, T. E., 1980. Fed. Proc. 39, 1049 (abstract).
15. Henson, P. M., Zanolari, B., Schwartzman, N. A. & Hong, S. R., 1978. J. Immunol. 121: 851–856.
16. Nelson, R. V., McCormack, R. T., Fiegel, V. D., Herron, M., Simmons, R. L. & Quie, P. G., 1979. Infect. Immun. 23: 282–286.
17. Gallin, J. I., Wright, D. G. & Schiffmann, E., 1978. J. Clin. Invest. 62: 1364–1374.
18. Becker, E. L. & Showell, H. J., 1974. J. Immunol. 112: 2055–2062.

Received December 8, 1980.

Muramyl peptides
Chemical structure, biological activity and mechanism of action

Arlette Adam[1], Jean-François Petit[1], Pierre Lefrancier[2] and Edgar Lederer[1,3]

[1] *Institut de Biochimie, Université de Paris-Sud, 91405 Orsay*

[2] *Institut CHOAY 92120 Montrouge*

[3] *Laboratoire de Biochimie, C.N.R.S. 91190 Gif-sur-Yvette*

Summary

The first two sections give a historical introduction describing the identification of a muramyl-dipeptide as the smallest adjuvant active moiety capable of replacing whole killed mycobacterial cells in Freund's complete adjuvant. This molecule, MDP (for muramyl-dipeptide = N-acetyl-muramyl-L-alanyl-D-iso-glutamine) was synthesized and shown to be fully active.

The third section describes in detail the various derivatives and analogues of MDP which have been synthesized and characterizes their main properties, such as adjuvant activity, stimulation of non-specific resistance to infections and pyrogenicity.

The fourth section gives a summary of the biological properties of MDP and its derivatives; these are reported under the following headings:

– Adjuvant and related activities (qualitative changes in antibody synthesis, immunogenicity of MDP, autoimmune responses, MDP as adjuvant for vaccines, induction of cell mediated immunity)

– Modulation of host resistance to tumours and to infections (antitumour activity of MDP and its derivatives, non-specific resistance to infection)

– Mitogenic activity, polyclonal activation, pyrogenicity of MDP.

The fifth section 'Target cells and mechanism of action' relates *in vivo* and *in vitro* experiments and the action of MDP on macrophages. Finally, the metabolic fate of MDP and derivatives is discussed.

The conclusions of this review try to foresee the future applications of MDP and its derivatives, as components of vaccines, as anticancer and antiparasitic agents and ends with a word of caution concerning the possible side effects of the intensive use of immunomodulators in human beings.

1. Introduction

The recognition of the immunomodulating properties of peptidoglycans and soluble peptidoglycan fragments is derived from work of the late twenties, showing that living mycobacteria are endowed with adjuvant activity, i.e. that they can increase the immune response of a host to foreign antigens. It was later proved that killed mycobacteria are also active, in an emulsion of water in mineral oil; in the well known Freund's Complete Adjuvant (1) (FCA), the mycobacterial cells are suspended in the oil phase, the antigen is dissolved, or suspended in saline and both phases are emulsified with an emulsifier: this FCA leads to an increase of the humoral response (circulating antibodies) towards the antigen and induces cellular immunity to the antigen, which can be measured by a classical test of delayed hypersensitivity, such as the skin test. If mycobacteria are omitted in the water in oil emulsion containing the antigen (Freund's Incomplete Adjuvant or FIA) one obtains a much smaller increase of the humoral response and no induction of cellular immunity towards the antigen.

Molecular and Cellular Biochemistry 41, 27–47 (1981).   0300-8177/81/0041-0027/$04.20.

Thus, the mycobacterial cell contains one or several components endowed with immunostimulant activity.

Investigations during the last decades led to the identification and structural elucidation of the adjuvant active structure, which is able to replace mycobacterial cells in FCA: the peptidoglycan (Fig. 1) which is active either intact, i.e. in the form of an insoluble macromolecule, or in the form of soluble fragments usually obtained by enzymatic hydrolysis of native peptidoglycan (2, 3).

We shall see that a simple glycodipeptide N-acetyl-muramyl-L-alanyl-D-isoglutamine(MDP1) is the smallest adjuvant active peptidoglycan fragment.

2. From Freund's adjuvant to MDP

The first step in identifying the adjuvant active structure of killed mycobacteria was the finding of White *et al.* (4), that the chloroform soluble wax D of *M. tuberculosis* and of *M. kansasii* is adjuvant

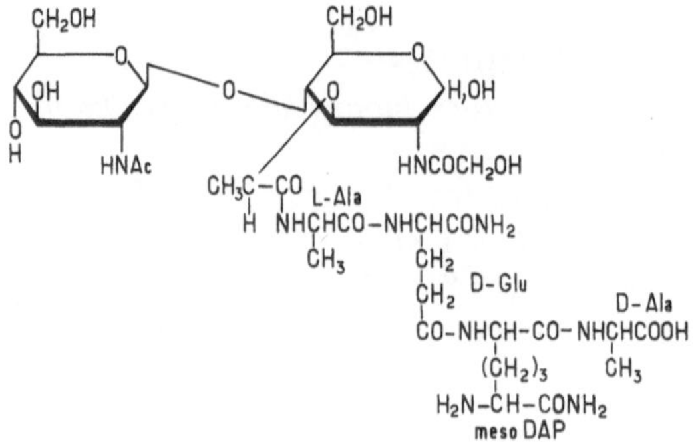

Fig. 1. Monomer of the peptidoglycan of Mycobacteria; N-acetyl D-glucosamine is linked to N-glycolyl muramic acid which carries a diamidated tetrapeptide L-Ala-D-isoGln-meso DAP-D-Ala (2, 3).

active. The chemical structure of these wax D fractions (Fig. 2 in a frame ....) was shown to resemble closely that of purified cell walls (which are also adjuvants, 5, 6), both containing an arabinogalactan esterified by mycolic acids and linked to a peptidoglycan (Fig. 2).

In 1972 it was reported that a lysozyme digest of purified cell walls of *M. smegmatis* contains hydrosoluble adjuvant active fractions. WSA (water

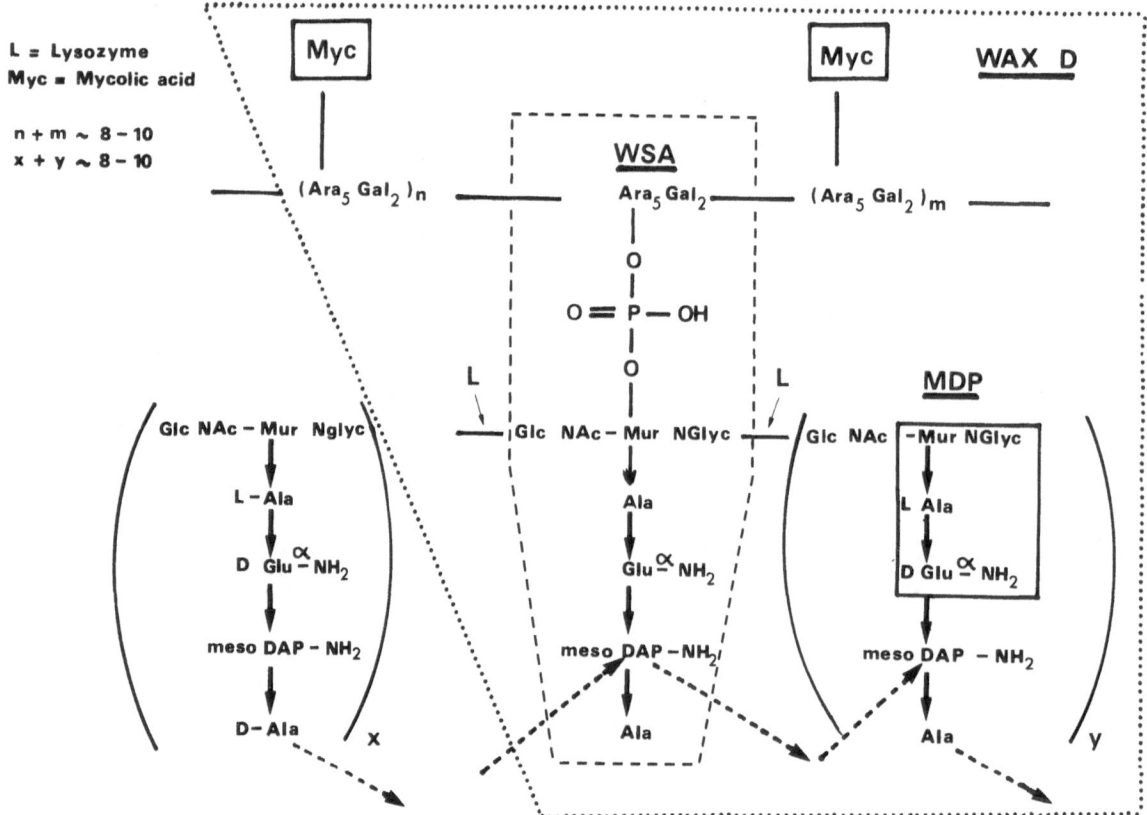

Fig. 2. Simplified scheme of the mycobacterial cell wall and its adjuvant active derivatives: Wax D, WSA, MDP (slightly modified from 3).

soluble adjuvant) (Fig. 2 in a frame - - - - - - -) was purified by filtration through Sephadex G 75 and found to have a molecular weight of approximately 20 000 dalton; it contains an arabinogalactan covalently linked to a fragment of the peptidoglycan (6).

Similar products were obtained by Migliore and Jollès (7) by aqueous extraction of delipidated cells of *M. tuberculosis* followed by ammonium sulfate precipitation and DEAE cellulose chromatography and by Hiu (8) by hydrogenolysis of delipidated cells of BCG suspended in a mixture of ethanol and glacial acetic acid.

By lysozyme treatment of delipidated cells of *M. smegmatis* instead of purified cell walls, followed by gel filtration Adam *et al.* (9), obtained a higher yield of a 'neo-WSA', similar to WSA but containing some non-peptidoglycan amino acids. Using delipidated cells of *Nocardia opaca* a fraction was isolated, which, besides cell wall constituents, contains products originating probably from the cytoplasmic membrane (10, 11), which have a potent mitogenic activity for B lymphocytes of various species including man (12) and are inducers of circulating interferon in mice (13) and hamsters (14).

Adjuvant active products containing neutral sugars covalently linked to a peptidoglycan fragment were also obtained by Stewart-Tull *et al.* (15), from culture filtrates of *M. tuberculosis* strain DT and by Kumazawa *et al.* (16), from *M. tuberculosis* strain Aoyama B using the hydrogenolysis technique of Hiu (8).

It was then shown that the neutral sugars are not necessary for adjuvant activity: indeed, purified insoluble peptidoglycans, from various Gram-negative organisms (17, 18) and from mycobacteria (19) are active, as well as soluble peptidoglycan fragments devoid of neutral sugars, obtained from mycobacteria by various ways (19).

Then it was shown that the repeating subunit of most peptidoglycans was able to replace mycobacteria in FCA; the third amino-acid can be *meso*-DAP as in mycobacteria (20), *Lactobacillus plantarum* (21) and Gram-negative species (18, 22), or lysine, as for instance in *Staphylococcus aureus* (21).

The absence of an amide on the γ-carboxyl group of D-Glu, which occurs in Gram-negative peptidoglycans, does not diminish the activity; the amino group of the muramic acid residue can be substituted by a glycolyl group, as in *Mycobacteria* (23, 24), or by an acetyl group as in most other genera, without change in activity.

The terminal D-alanine of the disaccharide tetrapeptide is not essential, as shown by the activity of the disaccharide tripeptides Glc-Nac-Mur-Nacyl-L-Ala-D-Glu-mesoDAP of *E. coli* or *M. smegmatis* (25). The glycopeptide structure is, however, necessary as the free disaccharide or the free tri- or tetrapeptides alone, obtained by enzymatic hydrolysis of the monomeric peptidoglycan, are inactive (25).

Finally, the N-acetylglucosamine group was shown not to be critical for activity: the N-acetyl-muramyl-tripeptides Mur-Nac-L-Ala-D-Glu-meso DAP (25, 26) and -pentapeptides Mur-Nac-L-Ala-D-Glu-mesoDAP-D-Ala-D-Ala (27) obtained by hydrolysis of the UDP-N-acetylmuramyl peptides are adjuvant, whether they contain L-Lys (as in *Staphylococcus epidermidis*) or *meso* DAP (as in *Bacillus megaterium*). Even, the third amino acid can be removed, the synthetic N-acetyl-muramyl-L-Ala-D-iso-Gln, or MDP *l* being active (28); amidation of the α-carboxyl group of Glu is necessary in FIA in the guinea pig test, but not in other tests (see below). N-acetyl-muramyl-L-Ala is inactive: MDP *l* is thus the minimal adjuvant active molecule of this series (25, 29).

MDP

The peptidoglycan being an ubiquitous structure in the procaryotic world, one can wonder, why the killed cells of only a few genera of bacteria, essentially *Mycobacteria* and *Nocardia,* are active in Freund's adjuvant (1). The reason is still not fully understood, the most probable interpretation being, that in most other bacteria the external constituents of the cell enveloppe (capsule, external membrane, teichoic acids, etc.) prevent the contact between the peptidoglycan and the immune cells. The isolated peptidoglycan of all the Gram negatives studied are indeed adjuvant (17), as are purified cell walls of most of the 21 Gram + species studied by Kotani *et al.* (30), the exception being *Micrococcus lysodeikticus,* and a few other Gram positive species (31). In the case of *M. lysodeikticus,* the reason of the inactivity might well be the substitution of the γ-carboxyl group of the D-Glu residue of the peptidoglycan by glycine: in the synthetic products (see Table 3, *59, 60*) such a substitution leads indeed to inactive compounds.

3. Synthetic muramyl peptides. MDP and derivatives

As discussed above a synthetic glycodipeptide, structurally derived from the bacterial cell wall: N-acetyl-muramyl-L-alanyl-D-isoglutamine (MDP) *1* can replace whole mycobacterial cells in Freund's complete adjuvant, for increasing the titre of antibodies and for producing delayed type hypersensitivity in guinea pigs (25, 28, 29).

Several hundred analogs of MDP have been synthesized and their immunomodulating properties and cellular mechanism of action studied. For reviews see (32–35).

In the following we describe in brief, the principal types of MDP derivatives, our present knowledge of structure-activity relationships and various aspects of the biological activities of MDP and its derivatives, both *in vivo* and *in vitro.* We cannot mention here the methods used for the synthesis of MDP and its derivatives; for leading references see (28, 36–39).

The major biological properties of MDP will be described in section 4.

Tables 1–5 give a list of the principal structural types of MDP derivatives with results concerning antibody production, delayed hypersensitivity and antibacterial activity.

Table I shows modifications of the carbohydrate

moiety of MDP (*1*). The intact muramic acid structure seems necessary for full adjuvant and anti-infectious activities (*2, 3, 4, 6* are inactive). Some modifications of the carbohydrate moiety lead, however, to active molecules: *nor*-MDP *7* (in which the methyl group of the muramyl side chain is replaced by H) is less active in several tests than MDP, but less pyrogenic.

Structural modifications of the C-2 substituent of the muramyl moiety of MDP have been described: the D-*manno* isomer *5* and the free amino compound *12* have stronger adjuvant activity than MDP (in oil), the D-gluco analogue *14* is also strongly active, whereas the 2-deoxy-D-*arabino*-hexose analogue *13* is inactive.

The primary hydroxyl group at the C-6 position can be acylated (*19, 20, 21, 22*) or replaced by an amino or an acylamino group (*17, 18*) without loss of adjuvant activity. The 6-deoxy- (*15*) and 6-deshydroxymethyl analogues (*16*) have, however, no adjuvant effect, but show antitumour activity.

α- or β-glycosidation of MDP decreases the adjuvanticity only on the humoral immune response (*11*).

Some disaccharide dipeptides (*8, 9*) have been prepared and seem to be more active than MDP. The enzymatically prepared disaccharide of the cell wall of *Lactobacilli* has been coupled to the dipeptide L-alanyl-D-isoglutamine; this disaccharide dipeptide *8* is as active as MDP against Sarcoma 180 and 3 to 7 times more adjuvant active than MDP (in saline); it has also been prepared by total synthesis.

It is possible that larger peptidoglycan fragments such as tetra-to hexasaccharides of di-to pentapeptides might have greater activity in certain tests; this is the case, for example, for arthritogenicity (55); they are, however, difficult to synthesize.

Table 2 shows MDP analogues in which the L-alanine residue is replaced by various aminoacid residues. For the expression of adjuvant activity L-alanine can be replaced by most of the L-aminoacids (or glycine): L-serine (*27*), L-valine (*32*) and L-α-aminobutyric acid (*33*) even seem to cause an enhanced adjuvant effect. The diastereo-isomeric N-acetyl-muramyl-D-alanyl-D-isoglutamine (MDP D-D) (*24*) is not only inactive, but under certain experimental conditions, it is an antagonist to MDP (44) and immunosuppressive (40).

N-methylation of L-Ala leads to a nearly apyro-

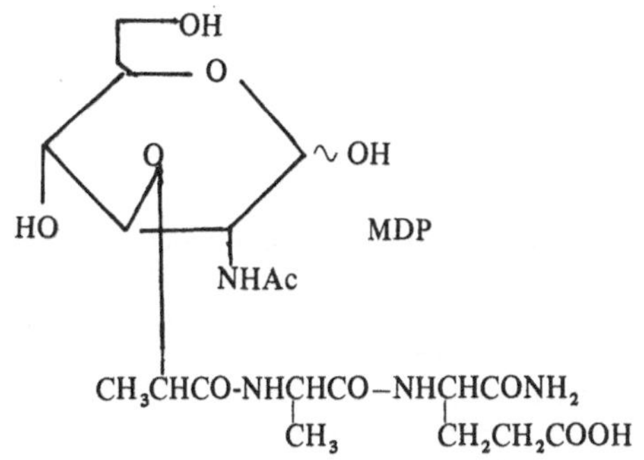

$CH_3CHCO\text{-}NHCHCO\text{-}NHCHCONH_2$

MDP

Table 1. Modifications of the N-acetyl-muramyl moiety.

		Adjuvant activity[1]					Anti-infectious activity[2]	
		Saline		Water-in-oil				
		Ab	refs.	Ab	DHS	refs.	Refs.	
1	MurNAc-L-Ala-D-isoGln[3]	+	(40, 41)	+	+	(25, 28, 41, 44, 53)	+	(51)
2	H-L-Ala-D-isoGln			-	-	(29, 42)		
3	Lactyl-L-Ala-D-isoGln			-	-	(43, 44)		
4	Muraminitol-NAc-L-Ala-D-isoGln			-	-	(44)		
5	D-*manno*-MurNAc-L-Ala-D-isoGln				+	(45)		
6	*iso*-MurNAc-	-	(39a)	-	-	(39a)		
7	*nor*-MurNAc-	+	(39a)	+	+	(39a, 44)	+	(39a)
8	β-GlcNAc(1 → 4)MurNAc-	+	(46)	+	+	(47)		
9	β-MurNAc(1 → 4)(GlcNAc)-	+	(48)					
10	β-Glc(1 → 4)MurNAc-				+/-	(45)		
11	1-O-α(β)-alkyl or arylMurNAc-[4]		(39a, 50)	-	+	(39a, 49, 50)	-	(39a, 51)
12	Mur-L-Ala-D-isoGln	-			+	(52)		
13	2-deamino-Mur-	-	(39a)	-	-	(39a, 52)	-	(39a)
14	2-hydroxy-2-deamino-Mur-				+	(52)		
15	6-deoxy-MurNAc-L-Ala-D-isoGln				-	(53)		
16	6-deshydroxymethyl-MurNAc-				-	(53)		
17	6-amino-6-deoxy-MurNAc- X [5]				+	(53)		
18	6-acetyl-amino-6-deoxy-MurNAc-L-Ala-D-isoGln				+	(53)		
19	6-O-acetyl-MurNAc-			+	+	(54)		
20	6-O-butyryl-MurNAc-			+	+	(54)		
21	6-O-succinyl-MurNAc-	+	(39a)	+	+	(39a)	+	(39a)
22	4,6-di-O-acetyl-MurNAc-	+	(32)	+	+	(32, 54)	+	(32)

[1] Administration with antigen to mice in saline, or to guinea pigs in a water-in-oil-emulsion.

[2] Administration to mice, one day before *K. pneumoniae* challenge.

[3] Abbreviations used for the carbohydrate moiety.

    MurNAc: 2-acetylamido-2-deoxy-3-O(D-carboxyethyl)-D-glucopyranose

    nor Mur-NAc: 2-acetamido-2-deoxy-3-O(D-carboxymethyl)-D-glucopyranose

    iso Mur-NAc: 2-acetamido-2-deoxy-4-O(D-carboxyethyl)-D-glucopyranose

    D-*manno* Mur-NAc: 2-acetamido-2-deoxy-3-O(D-carboxyethyl)-D-mannopyranose

    GlcNAc: 2-acetamido-2-deoxy-D-glucopyranose

    2-deamino-Mur: 2-deoxy-3-O(D-carboxyethyl)-D-arabinopyranose

2-hydroxy-2-deamino MurNAc: 3-O-(D-carboxyethyl)-D-glucopyranose

    6-deoxy-MurNAc: 2-acetamido-2,6-dideoxy-3-O(D-carboxyethyl)-D-glucopyranose

6-deshydroxymethyl-MurNAc: 2-acetamido-2-deoxy-3-O-(D-carboxyethyl)-O-xylopyranose

[4] α and β-methyl glycosides, β-*p*-aminophenylglycoside

[5] x = L-Ala, L-Val, L-Ser

genic adjuvant active molecule (*34*) but which is not anti-infectious (34).

Table 3 shows modifications of the second aminoacid residue (D-Glu-) of MDP; this part of the molecule has a particular importance for the expression of both adjuvant and anti-infectious activities. Three structural features are essential: stereo-chemistry (the L-isomer *43* is inactive), length of the side chain (the Asp analog *42* is inactive) functionality of the α- and γ-carboxylgroups (*40, 41*). The influence of the substitution of the latter on the activity has been studied. If the α-carboxyl

OH
O
O
HO
NHAc
MDP
∿ OH

CH₃CHCO-NHCHCO—NHCHCONH₂
     CH₃   CH₂CH₂COOH

Table 2. Modifications of the first amino acid residue of MDP.

| | | Adjuvant activity[1] | | | | | Anti-infectious activity[1] | |
| | | Saline | | Water-in-oil | | | | |
		Ab	Refs.	Ab	DHS	Refs.		Refs.
1 MurNAc-L-Ala——→D-isoGln	*1*	+	(40, 41)	+	+	(25, 28, 29, 41, 44, 53)	+	(51)
23 ———D-isoGln —— OH	*23*			–	–	(29)		
24 ———D-Ala ——— D-isoGln	*24*	–	(40)	–	–	(40, 44, 56)	–	(51, 57)
25 ———β-Ala———	*25*	–	(39a)	–	–	(39a)	–	(39a)
26 ———Gly———	*26*	+/–	(40)	+/–	+/–	(44, 56)	–	(51)
27 ———L-Ser———	*27*	+	(40)	+	+	(44, 56)	–	(51)
28 ———L-Leu———	*28*			–	–	(156)		
29 ———L-Tyr———	*29*			+	+	(156)		
30 ———L-Pro———	*30*			+	+	(156)		
31 ———L-Thr———	*31*			+	+	(156)		
32 ———L-Val———	*32*			+	+	(156)		
33 ———L-α-Abu———	*33*	+	(40)	+	+	(33, 156)		
34 ———N-Me-L-Ala———	*34*	+	(34)	+	+	(34)	–	(34)
35 ———N-Me-Gly———	*35*	–	(39a)	–	–	(39a)	–	(39a)
36 ———N-Me-L-Leu———	*36*	–	(39a)	–	–	(39a)	–	(39a)

[1] for footnotes, see Table 1.

function is amidated, many substitutions can be made on the γ-carboxyl group without loss of adjuvant activity: amidation (*44*), esterification (*46*), substitution by free or esterified aminoacid residue (*47, 48, 49, 50*) or peptides (*51, 52, 53*), but not methylamidation (*45*).

The requirements for anti-infectious activity are, however, more stringent: compounds bearing an amide (*44*), a methylamide group (*45*) a methyl ester (*46*) or a D-alanine residue at the C-terminal end (*47, 51*) are inactive as well as α-methylamide analogues (*54, 55, 56*). The α-glycine analogues (*57, 58, 59*) have no activity. Compounds bearing a free α-carboxyl function (*57, 58*) do not have any adjuvant activity when administered in water-in-oil emulsion.

The α-ester-analogues (*63, 64, 65, 66, 67, 68*) are mostly active. Among them, the most interesting is MurNac-L-Ala-D-Gln-O-nC₄H₉ (*67*).

Table 4 shows a series of lipophilic derivatives of MDP, most of them especially active on the cellular immune response. They are of great interest since, Japanese authors' showed that 6-O-mycolyl-MDP (*79*), 6-O-nocardomycolyl-MDP (*76*) and 6-O-corynomycolyl MDP (*75*) have strong antitumour activity and are only weakly pyrogenic (61).

Takada *et al.* (66a), have prepared various 6-O-acyl derivatives of MDP with C_{18} to C_{48} linear, α-branched and α-branched, β-hydroxylated fatty acids and have shown that in particular 6-O-(2-tetradecylhexadecanoyl)-MDP and 6-O-(3-hydroxy-2-tetradecyl-octadecanoyl)-MDP exhibit stronger macrophage-stimulating effects than MDP. See also Pabst *et al.* (66b).

Another category of lipophilic MDP derivatives bears the lipid moiety at the C-terminal end of the peptide chain. One of the most lipophilic compounds, MDP-L-Ala-glycerol-mycolate (*73*) strongly stimulates nonspecific resistance against bacterial infections (67). The introduction of a second mycolyl group such as in *82, 83* does not seem to have any pharmacological advantage.

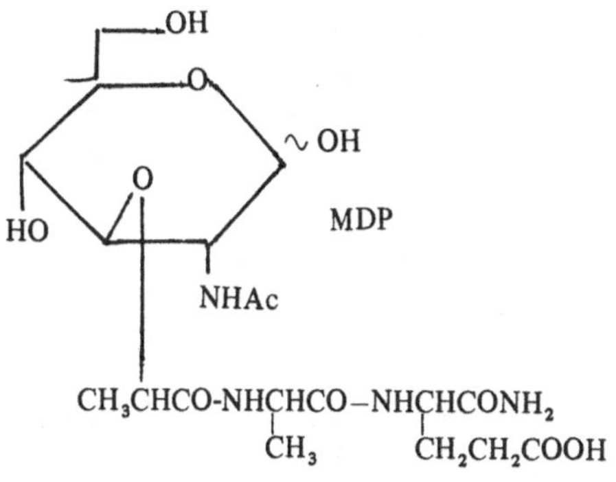

OH

O

~ OH

O

MDP

HO

NHAc

CH₃CHCO−NHCHCO−NHCHCONH₂

$CH_3CHCO-NHCHCO-NHCHCONH_2$

CH_3 CH_2CH_2COOH

Table 3. Modifications of the second aminoacid residue of MDP.

	Adjuvant activity[1]					Anti-infectious activity[2]
	Saline		Water-in-oil			
	Ab	Refs.	Ab	DHS	Refs.	Refs.
1 MurNAc–L-Ala–D-Glu-NH₂ ⌐————OH	+	(40, 41)	+	+	(25, 28, 29, 41, 44, 53)	+ (51)
37 MurNAc–L-Ala-OH	–	(32)	–	–	(29, 40, 42)	– (39)
a) Replacement of D-Glu by other aminoacids						
38 MurNAc–L-Ala–D-Ala-OH			–	–	(29)	
39 ————————D-Ala-NH₂			–	–	(29)	
40 ————————D-Nle-NH₂	–	(58)	–	–	(58)	– (51)
41 ————————γ-Abu	–	(58)	–	–	(58)	– (51)
42 ————————D-Asp-NH₂	–	(58)	–	–	(58)	– (51)
43 ————————L-Glu-NH₂	–	(40)	–	–	(29, 44, 58)	– (39)
b) Substitution of D-Glu-α-NH₂						
44 MurNAc–L-Ala–D-GluNH₂ ⌐———NH₂	+	(58)	+	+	(58)	– (51)
45 ————————NHR (2)	–	(58)	–	–	(58–60)	– (51)
46 ————————OCH₃	+	(40)	+	+	(40, 44)	– (51)
47 ————————D-Ala	+	(39)	+	+	(39)	– (39)
48 ————————L-Ala	+	(39)	+	+	(39)	+ (39)
49 ————————L-Ala-OnC₄H₉	+	(39a)	+	+	(39a)	+ (39a)
50 ————————L-Lys	+	(32)	+	+	(29, 32)	+ (51)
51 ————————L-Lys-D-Ala	+	(58)	+	+	(32, 42, 58, 60)	– (51)
52 ————————L-Lys-L-Ala	+	(39)	+	+	(39)	+ (39)
53 ————————L-Lys(Ac)-D-Ala-NH₂				+	(59)	
54 MurNAc–L-Ala–D-Glu-NHCH₃ ⌐———OH	+	(40)	+	+	(40)	– (51)
55 ————————NH₂	+	(39a)	+	+	(39a)	– (39a)
56 ————————NHCH₃	–	(58)	–	–	(58)	– (51)
57 MurNAc–L-Ala–D-Glu-OH ⌐———OH	+	(40, 41)	–	–	(29, 40, 41, 44)	+ (51)
58 ————————NH₂	+	(32)	–	–	(29, 32)	+ (39a)
59 MurNAc–L-Ala–D-Glu-Gly ⌐———L-Lys-D-Ala			–	–	(60)	
60 MurNAc–L-Ala–D-Glu-Gly	–	(58)	–	–	(58, 60)	– (39a)
61 ————————Gly-NH₂	+	(39a)	+	+	(60)	– (39a)
62 ————————D-Ala-NH₂			+/–	+/–	(60)	
63 MurNAc–L-Ala–D-Glu-OCH₃ ⌐———OH	+	(32)	+	+	(32)	+ (39a)
64 ————————NH₂	+	(39a)	+	+	(39a)	– (39a)
65 ————————OCH₃	+	(40)	+	+	(40, 44)	+ (51)
66 MurNAc–L-Ala–D-Glu-OnC₄H₉ ⌐———OH	+	(39a)	+	+	(39a)	+ (39a)
67 ————————NH₂	+	(39a)	+	+	(39a)	+ (39a)
68 ————————OCH₃	+	(39a)	+	+	(39a)	+ (39a)

[2] R = -methyl, *n*-hexyl, stearyl

Table 4. Lipophilic MDP derivatives.

		Adjuvant activity[1]					Anti-infectio activity[1]	
		Saline		Water-in-oil				
		Ab	Refs.	Ab	DHS	Refs.		Refs.
69	MurNAc–L–Ala–D–isoGln–L–Lys[ε]–mycolyl				+	(63)		
70	——————————————— NH–(CH$_2$)$_2$O–mycolate				+	(63)		
71	——————————————— L–Ala–O–n–C$_{10}$H$_{21}$	+	(39)	+	+	(39)	+	(39)
72	——————————————— L–Ala–O–n–C$_{20}$H$_{41}$	+	(39)	+	+/–	(39)	+	(39)
73	——————————————— L–Ala–glycerol–mycolate	+	(39)	+/–	+	(39)	+	(39)
74	6-O-acyl-MurNAc–L–Ala–D–isoGln[2]				+	+	(54, 65)	
75	6-O-corynomycolyl-MurNAc ———————				+	(61)		
76	6-O-nocardomycolyl-MurNAc ———————				+	(61)		
77	6-O-stearoylamino-6-deoxy-MurNAc ———————				+	(53)		
78	6-O-Mycolylamino-6-deoxy-MurNAc ———————				–	(53)		
79	6-O-Mycolyl-MurNAc–L–Ala–D–isoGln			+/–	+	(61, 62, 64)		
80	———————————— Gly ————————			+/–	–	(64)		
81	———————————— L–Ser————————			+	+	(64)		
82	———————————— L–Ala–D–isoGln–L–Lys[ε]–(Mycolyl)				+	(63)		
83	6-O-mycolyl-MurNAc–L–Ala–D–isoGln–L–Lys				+	(63)		
	6-O-mycolyl-MurNAc——————————‖							
84	6-O-quinonyl-MurNAc–L–Val–D–isoGln				+	(66)		
85	6-O-quinonyl-MurNAc- X ————OMe[3]				+	(66)		

[1] See footnotes of Table 1.
[2] 6-O-acyl: 6-O-octanoyl, 6-O-lauroyl, 6-O-stearoyl, 6-O-doc sanoyl.
[3] X = Val, Ser, Thr.

Various MDP esters of acids derived from ubiquinones have also been prepared (*84, 85*) and have strong antitumour activity (66).

Table 5 shows a series of *peptidolipids*. The 'desmuramyl' derivative of MDP-L-Ala-glycerol-mycolate, i.e. L-Ala-D-isoGln-L-Ala-glycerol-mycolate ('triglymyc') (*92*) is just as active as the MurNac derivative in stimulating nonspecific antibacterial resistance; it is, however, entirely inactive as an adjuvant (67). This shows that in this series the MurNac moiety plays an essential role in stimulating humoral antibody production. The L-Ala-D-Glu moiety seems optimal in triglymyc as in MDP itself, thus suggesting that there exists a specific receptor (on the macrophage?) for this structure. The replacement of the dipeptide by L-Ala-L-isoGln (*93*), L-Ala-D-isoAsn (*94*), β-Ala-D-isoGln (*95*), or L-Ala-D-Gln *97* leads to much less active products. The length of the lipid chain is also important: the mycolate gives optimum activity; a 'corynomycolate' *91* (with a synthetic isomer of the C$_{32}$ corynomycolic acid) is only weakly active (67).

Migliore-Samour *et al.* (68), have recently prepared an active peptidolipid by *N*-laurylation of an adjuvant inactive cell wall tetrapeptide isolated from *Streptomyces*. The synthetic N[2]-[N-(*N*-lauroyl-L-alanyl)-γ-D-glutamyl] N[6]-(glycyl)-DD,LL diamino-2,6 pimelamic acid *100* stimulates delayed hypersensitivity to ovalbumin (in FIA) and protects mice against infection by *Listeria monocytogenes*. This new type of compound (a mixture of diastereoisomers) increases circulating antibodies despite the absence of the MurNac moiety.

Mašek *et al.* (59), have reported that the nonapeptide L-Ala-D-isoGln-L-Lys(Ac)-D-Ala-(Gly)$_5$-OMe (*99*) stimulates production of delayed hypersensitivity when injected in FIA to mice with ovalbumin as antigen; this report has not yet been confirmed.

Oligomers and polymers

Mainly due to its small molecular weight MDP is not antigenic (41), it does not react with antipeptidoglycan antibodies and does not sensitize to

Table 5. Peptidolipids.

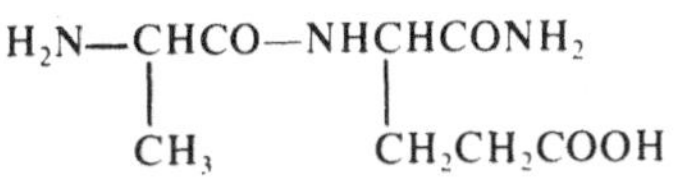

$H_2N-CHCO-NHCHCONH_2$
    | |
    CH_3 CH_2CH_2COOH

| | Adjuvant activity[1] | | | | | Anti-infectious activity[1] |
| | Saline | | Water-in-oil | | | |
	Ab	Refs.	Ab	DHS	Refs.	Refs.
86 L-Ala-D-isoGln	−	(67)	−		(29, 42, 67)	− (67)
87 L-Ala-D-isoGln-L-Ala-On$C_{10}H_{21}$	−	(67)	−		(67)	− (67)
88 ————L-Ala-On$C_{15}H_{31}$	−	(67)	−		(67)	− (67)
89 ————L-Ala-glycerol-stearate	−	(67)	−		(67)	− (67)
90 ————L-Ala-glycerol-distearate	−	(67)	−		(67)	− (67)
91 ————L-Ala-glycerol-corynomycolate	−	(67)	−		(67)	− (67)
92 ————L-Ala-glycerol-mycolate	−	(67)	−		(67)	+ (67)
93 ————L-isoGln-L-Ala-glycerol-mycolate	−	(67)	−		(67)	− (67)
94 ————D-isoAsn-L-Ala-glycerol-mycolate	−	(67)	−		(67)	− (67)
95 β-Ala-D-isoGln-L-Ala-glycerol-mycolate	−	(67)	−		(67)	− (67)
96 D-isoGln-L-Ala-glycerol-mycolate	−	(67)	−		(67)	+ (67)
97 L-Ala-D-Gln-L-Ala-glycerol-mycolate	−	(67)	−		(67)	
98 ————L-Lys(Ac)-D-Ala-NH_2				−	(59)	
99 ————L-Lys-D-D-Ala-(Gly)$_5$-OMe				+	(59)	
100 Lauroyl-L-Ala-D-Glu-α ⌐ LL,DD-DAP ⌐ Gly—NH_2			+	+	(68)	+ (68)[2]

[1] See footnotes of Table 1.

[2] Protection against *L. monocytogenes* infection.

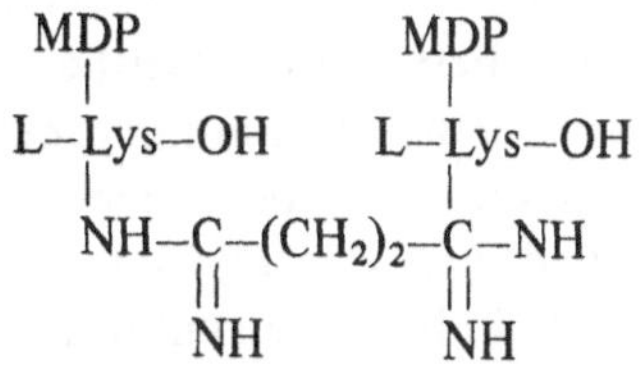

```
        MDP              MDP
         |                |
      L–Lys–OH         L–Lys–OH
         |                |
      NH–C–(CH₂)₂–C–NH
         ||              ||
         NH              NH

              101
```

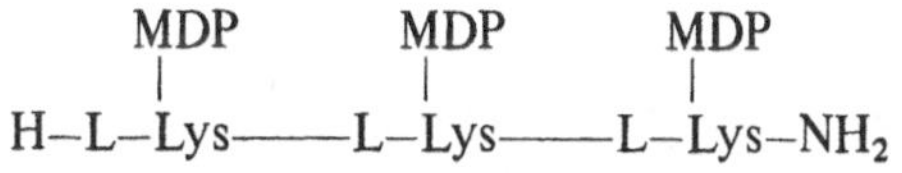

```
     MDP        MDP        MDP
      |          |          |
   H–L–Lys——L–Lys——L–Lys–NH₂

              102
```

tuberculin (69). It is, however, very rapidly excreted in the urine: more than 50% after 30 min, more than 90% after 2 h in mice (70). Moreover, MDP is not active in some systems where larger peptidoglycan fragments are active (71).

In view of increasing the life time of MDP in the organism, oligomers and polymers of MDP have been prepared, dimers such as *101* and trimers such as *102* (39) and a β-D-p-aminophenyl glycoside of MDP, cross-linked with glutaraldehyde. The glycoside had lost most of the activities of MDP, whereas, some of these were recovered and even increased after cross-linking (50).

MDP on carriers

Other possibilities of increasing the size and efficiency of MDP have been tried. Reichert *et al.* (72), have conjugated MDP to several protein carriers via carbodiimide or phenylisothiocyanate. Chedid *et al.* (73), have combined MDP with multi-(poly-DL-alanyl-poly-L-lysine) and found, that, indeed, this 'macromolecularization' does not increase adjuvant activity but potentiates the anti-infectious activity of MDP 100 fold, but also its pyrogenicity. The 'inactive' (or even anti-adjuvant and immunosuppressive) stereoisomer of MDP

Table 6. Adjuvant activity of MDP measured by antibody production (Ab), delayed hypersensitivity (DHS) or increased production of plaque forming cells (PFC).

Test antigen	Species	Response		References
Ovalbumin in FIA	guinea pig	Ab	DHS	(25, 28, 29, 41, 76a)
Ovalbumin in FIA	rat		DHS	(77)
ABA-N-acetyl tyrosine in FIA	guinea pig		DHS	(41, 43)
Bacterial α Amylase in FIA	mouse	Ab		(43)
	guinea pig	Ab	DHS	(42)
DNP ficoll in saline	mouse	PFC		(43)
Bovine serum albumin	mouse	Ab in primary and secondary response		(40–41)
DNP-GAT in FIA or saline	mouse	Ab		(78)
SRBC	mouse	PFC		(79, 80, 81)

(MDP-DD) *24* coupled to that carrier, gives a molecule which is much less pyrogenic and is not an adjuvant, but which increases strongly non specific, antibacterial resistance. Moreover, it does not sensitize, nor induce an immune response to the glycopeptide moiety. A recent patent of Ciba-Geigy describes antigen-MDP and antigen-carrier-MDP compounds (74).

Sela and Mozes (75) have stressed the future of combining a synthetic antigen via a synthetic carrier with a synthetic adjuvant.

4. Biological properties of MDP and its derivatives

We shall first discuss the biological effects of MDP and its derivatives and then the mechanism of action of MDP.

1. Adjuvant and related activities

As mentioned above, adjuvant activity is a common property of peptidoglycan derived products. In screening experiments adjuvant activity is usually measured with compounds injected in FIA; under these conditions they increase both cellular and humoral immunity; MDP and some derivatives are also able to increase antibody levels when injected in saline (41). MDP has been used as an adjuvant in various species and with different antigens (Table 6).

1a. Qualitative changes in antibody synthesis

Adjuvants that enhance the antibody response,

may induce qualitative changes in immunoglobulins. The injection of a soluble antigen in mineral oil in guinea pigs induces first IgM globulins (after 1 week), then IgG_1 (after 3 weeks) but the addition of mycobacteria in FCA induces the synthesis of IgG_2 globulins (76) which fix complement and have cytophilic properties; their presence has been correlated with delayed hypersensitivity in immunized animals. MDP in FIA was also shown to induce the synthesis of IgG_2 in guinea pigs (29, 76a, 82), both in the primary and the secondary response; a good correlation between the synthesis of IgG_2 and the presence of cellular immunity, detected by cutaneous reaction or appearance of MIF could be established (76a). Only adjuvant active MDP derivatives induce the synthesis of IgG_2 globulins (76a).

MDP was shown to be active in saline in mice (40, 41) essentially for the secondary response; high levels of reaginic antibodies are synthesized (83, 84); Heymer *et al.* (85) have observed an anaphylactic reaction or even systemic anaphylaxis in the secondary response of mice preimmunized with MDP; they correlated these findings with high levels of IgG_1, which are known to mediate anaphylactic reactions in mice; a decrease of IgG_2 globulin was noted (78, 82).

The covalent linkage between adjuvant and antigen, can greatly modify the immunoglobulin pattern. Kishimoto *et al.* (86) have shown that the IgE antibody synthesis which is regulated by isotype specific regulatory T cells, can be inhibited by the preadministration of the allergen coupled to mycobacterial cells; the preimmunization with DNP or

ovalbumin covalently linked to 6-O-Mycoloyl-
MDP-Lys or also to MDP-Lys, injected in FIA,
prevents the further appearance of DNP or ovalbumin IgE antibodies when animals are challenged
with DNP-ovalbumin or ovalbumin; IgG antibody
levels are unchanged; this suppressive effect is
mediated by radiosensitive T cells induced by the
previous injection of the hapten- or antigen-cou-
pled adjuvant.

1b. Immunogenicity of MDP

MDP is not immunogenic; its injection with
either FIA or FCA does not lead to any antibody or
delayed hypersensitivity reaction to MDP itself
(41). As MDP is part of the bacterial peptidoglycan
a search for cross reactions with naturally occurring
or experimentally induced anti-peptidoglycan
antibodies was made; there was no binding of MDP
(69), in agreement with the finding that the
immunodominant determinant of bacterial peptidoglycans is the pentapeptide: L-Ala-D-isoGln-L-
Lys-D-Ala-D-Ala (88). Recently (72) MDP was
coupled to carrier proteins; these conjugates could
elicit the synthesis of anti-MDP antibodies although
their adjuvant potentialities were maintained. Similarly, in contrast with MDP alone, MDP conjugated with multi-(poly-D-L-ala)-(poly-L-lys)
(MDP-AL) behaves as an elicitin and a sensitin,
when administered with FCA (89).

1c. Autoimmune responses

FCA is often used to induce autoimmune
responses, when injected with auto-antigens.
Although MDP injected in FIA does not induce
allergic polyarthritis in rats (90) it has been used to
study the induction of autoimmune diseases, such
as orchitis (91), uveo retinitis (92) or arthritis
(55, 93).

Nagao & Tanaka (93a) have reported, however,
that MDP does induce adjuvant arthritis in WKA
rats when injected in a water-in-oil emulsion prepared with Freund's incomplete adjuvant (Difco),
but not when emulsified with Drakeol and Arlacel
A. Experimental allergic encephalomyelitis was
particularly studied because it is a model of cell
mediated autoimmune disease; it can be induced,
when myelin basic protein is injected with FCA.
Nagai et al. (94, 95) showed that a nonapeptide
(H-Phe-Ser-Trp-Gly-Ala-Glu-Gly-Gln-Arg-OH) is
sufficient to induce the disease and that MDP is the

smallest compound able to replace mycobacteria in
FIA; moreover, with MDP, even the heptapeptide
(Trp-Gly-Ala-Glu-Gly-Gln-Arg-OH) could induce
the disease.

1d. MDP as adjuvant for vaccines

MDP is capable of improving the efficiency of
influenza virus vaccines in mice (96, 97); 6-O-acyl
derivatives of MDP in squalene emulsion are also
active in guinea pigs (97a).

Protective immunity against malaria was produced in Aotus monkeys and in macaques using
MDP or nor-MDP in FIA and Plasmodium
falciparum or P. knowlesi merozoites (98, 98a).
With 6-O-stearoyl-MDP in liposomes a full protection of Aotus monkeys was reported using mature
segmenters of P. falciparum as the antigen (99).

Partial immunity to Schistosoma mansoni in rats
has been produced by injection of MDP in FIA, in
presence of a schistosomal antigen (100). In these
experiments the main effect of MDP seems to be
the stimulation of non specific resistance; the
inclusion of an antigen does not significantly
increase the degree of immunity.

1e. Induction of cell mediated immunity

MDP injected with FIA has been shown to
induce a delayed state of hypersensitivity (DHS) to
heterologous proteins or ABA tyrosine (Table 6)
and to induce experimental allergic encephalitis in
animals injected with an encephalitogenic peptide
(94, 95); it has also been shown that antigen and
MDP as well as 6-O-mycolyl MDP included in
liposomes were effective in inducing DHS (54); 6-
O-mycolylated MDP derivatives are effective adjuvants for induction of a cell mediated immunity to
allogeneic cells, even when injected in saline (61).
Recently the effects of MDP on various aspects of
CMI such as T cell proliferative response, induction
of cytotoxic T cells etc. were studied and showed a
good correlation between in vivo and in vitro
experiments (101).

2. Modulation of host resistance to tumors and to infections

2a. Antitumour activity of MDP and its derivatives

Many bacterial products modulate host resistance (102). MDP itself does not seem to have any
antitumour effect. However, Japanese workers

Table 7. Tumor regression after treatment. Values are numbers of guinea pigs cured of dermal and metastatic tumors over numbers of animals treated. Data shown are pooled from two separate experiments. No cures were observed in animals treated with any muramyl dipeptide in the absence of trehalose dimycolate (107).

Synthetic muramyl dipeptide (150 μg) tested with trehalose dimycolate (150 μg)	Observed tumor regression
N-Acetylmuramyl-L-alanyl-D-isoglutamine	1/17
N-Acetylmuramyl-D-alanyl-D-isoglutamine	0/9
N-Acetyl-4,6-di-*O*-octanoylmuramyl-L-alanyl-D-isoglutamine	1/9
N-Acetyldesmethylmuramyl-L-alanyl-D-isoglutamine[1]	2/19
N-Acetylmuramyl-L-threonyl-D-isoglutamine	3/9
N-Acetylmuramyl-L-seryl-D-isoglutamine	10/17[2]
N-Acetyldesmethylmuramyl-L-valyl-D-isoglutamine[1]	10/17[2]
N-Glycolyldesmethylmuramyl-L-alanyl-D-isoglutamine	7/9[2]
N-Acetyl-4,6-di-*O*-octanoylmuramyl-L-valyl-D-isoglutamine	16/18[2]
N-Acetyldesmethylmuramyl-L-α-aminobutyryl-D-isoglutamine[1]	17/18[2]
Trehalose dimycolate alone (control)	0/17
Emulsion of oil, Tween 80, and phosphate-buffered saline (control)	0/14

[1] These are trivial names for 2-acetamido-2-deoxy-D-gluco-3-*O*-yl-acetyl dipeptides.

[2] Significantly different from the value for trehalose dimycolate-treated controls.

have described a whole series of 6-O-acyl MDP esters with high molecular weight natural mycolic acids, such as 6-O-corynomycolyl-MDP (*75*), 6-O-nocardomycolyl-MDP (*76*) and 6-O-mycolyl-MDP (*79*), which are active against various tumours, such as a 3-methylcholanthrene induced fibrosarcoma of the mouse (61, 64).

Compounds such as *82* and *83* with two mycolyl groups, or *84* and *85* with an ubiquinone derived lipophilic acid are also antitumour active (63, 66).

In vitro, MDP and some derivatives are able to increase the cytostatic activity of elicited macrophages against a syngeneic tumour: this will be discussed later.

The combined immunostimulation by MDP and its derivatives, with trehalose dimycolate, (TDM), (35, 102a) leading to regression of dermal tumours and metastases in guinea pigs has been recently described.

When line 10 tumour of strain 2 guinea pigs is inoculated intradermally it metastasizes in the draining lymphnodes (103). It has been shown previously that it regresses after intralesional injection of mycobacterial cell walls (104) or TDM + purified cell walls (105), or TDM + an endotoxin (106); these preparations, when emulsified in saline containing 0.2% Tween 80 and 0.75 to 10% mineral oil induce regression of the tumour itself and the lymphnode metastases, the cured animals being immune against a later challenge graft of the same tumour.

In this system, MDP can replace mycobacterial cell walls, or the endotoxin preparation, when it is

Table 8. Tumour regression induced in guinea pigs by intra-lesional injection of emulsified mixtures of MDP and TDM: Dose response (108).

Material injected[1] (mg)		No. of cured animals[2]/No. of animals tested (90 days)
MDP	TDM	
0.25	–	0/7
0.25	1	8/8
0.25	0.2	8/8
0.25	0.04	6/8
0.05	0.2	7/8
0.05	0.04	7/8
0.01	1	7/8
0.01	0.2	7/8
0.01	0.2	7/8
0.01	0.04	6/7
–	1	0/7

[1] Emulsions were prepared by ultrasonication and contained 2% squalane and 0.2% Tween.

[2] Complete disappearance of the dermal tumour, no clinical evidence of metastatic disease, and rejection of contralateral challenge (10^6 tumour cells) 2 months after the inoculation of the tumour transplant.

administered with TDM, or even with a synthetic lower homologue, C_{76}, in an oil in water emulsion.

McLaughlin et al. (107) (Table 7) use several MDP derivatives (150 μg) with 150 μg P_3 (a highly purified TDM preparation). Yarkoni et al. (108) (Table 8), obtain a large percentage of cures with doses of less than 100 μg MDP + 100 μg TDM, or C_{76}. Here again, all cured animals are resistant to a challenge dose with the same tumour. Yarkoni & Rapp (109) have studied in detail the nature and % of oil necessary for obtaining the best effect with cell walls, and later with MDP + TDM; squalane, or Drakeol are much better than squalene or hexadecane. In Table 8, 2% squalane + 0.2% Tween is used.

McLaughlin et al. (107), find MDP ineffective, whereas several analogues are highly effective (Table 7). The discrepancy between the results of both groups might be principally due to the difference in % of oil, McLaughlin et al. (107) using only 0.75% oil.

Combined immunostimulation by MDP (or a less pyrogenic derivative) with the synthetic C_{76} (which is not toxic and much less granulomagenic than TDM) (110), in 2% squalane, a slowly metabolizable hydrocarbon, opens good perspectives for clinical use.

2b. Non-specific resistance to infection

Natural resistance to infection can be increased by previous parenteral injection of immunostimulants such as BCG or LPS, 2 weeks or one day respectively before challenge with Klebsiella pneumoniae (111). In contrast, MDP can enhance the non-specific immunity even when injected i.v. or s.c. or given per os, at the time of or up to 4 days before K. pneumoniae infection (51). MDP is also able to protect neonate mice (112) although definitive protection associated with a greater destruction and blood clearance of bacteria can be obtained only in 8 day old mice. MDP can protect mice against strains of K. pneumoniae made resistant to antibiotics by plasmid transfer (111) and can restore the non-specific immunity of immunosupressed mice (113). Such animals (pretreated with cyclophosphamide) have a greatly increased susceptibility to various infections as well as a reduction of the efficacy of antibiotics; nor-MDP 7 was shown to counteract the effect of immunosuppressive treatment, 'indicating a potential clinical usefulness of

immunostimulation in immuno-compromised and infected human patients' (114). MDP can also increase the efficacy of antibiotics in experimental bacterial and fungal infections in mice (115, 116). In FIA MDP is also active against Listeria monocytogenes in nude mice (116a). MDP, nor-MDP and the α-aminobutyryl analog of MDP increase resistance of mice to Pseudomonas aeruginosa and Candida albicans (117).

The capacity of MDP to induce non specific immunity to K. pneumoniae was improved by polymerization of its β-D-p-amino-phenyl glycoside, which although adjuvant active in guinea pigs was not able to protect mice against infection; its cross linkage by glutaraldehyde gave, however, a product (molecular weight 6 000 daltons), which had a stronger protective activity than MDP itself (50). The coupling of MDP to a multipoly-(D,L-Ala)-poly(L-Lys) carrier was also shown to improve the property of enhancing non specific immunity; a protection can even be obtained against L. monocytogenes which is insensitive to MDP itself (73, 118).

Partial protection of mice against Trypanosoma cruzi has been obtained by implantation of 4 mg MDP in saline in an Alzet minipump (119) which releases 1 μl/h, during 7 days.

With a single dose of MDP (0.5 mg), enhanced resistance to T. cruzi was obtained only when the drug was injected 48, but not 24 or 2 h before infection. The authors concluded that their results prove the role of phagocytes in the induction of resistance to T. cruzi infection (119).

3. Mitogenic activity

MDP produces only a weak increase in ^{3}H thymidine uptake by splenocytes (120, 121, 122) as compared to LPS (a 3-fold increase instead of a 15-fold increase). MDP acts on splenocytes of nude mice (123) but has no effect on thymocytes (124). The mitogenic activity of MDP which is optimal at day 4 to 5 depends on experimental conditions and on animal strains: for example DBA/2 spleen cells can be stimulated by MDP but not $C_{57}BL/6$ cells (125). The study of various analogues in high responder strains shows a good correlation between adjuvant and mitogenic activity (122) but it can not be concluded that the immunoenhancing properties of MDP are directly related to its weak mitogenic activity and it was suggested that MDP requires a polymerized form for a full expression of the mitogenic activity on B cells (121).

4. Polyclonal activation

It has been suggested (126) that the adjuvant effect could be due to a stimulation of 'background' immunocytes leading to a generalized clonal expansion of committed lymphocytes. It has been shown (43, 120, 122, 127) indeed that MDP enhances the anti SRBC, PFC response in spleen cells in absence of any antigenic stimulation: this effect is comparable to the polyclonal activation obtained with LPS. This activity can be expressed against self or modified self antigen and can be found in macrophage depleted spleen cells or cultures from nu/nu spleen cells. This activity can result from an activation of B lymphocytes (128). Surprisingly, the LPS induced polyclonal activation can be inhibited by MDP, even when it is added 24 h after initiation of cultures, although the mitogen induced blast transformation is not affected (129).

5. Pyrogenicity of MDP

Pyrogenicity of MDP was first thought to be correlated with its adjuvant properties (130), only active compounds being able to elicit a febrile response in rabbits, which is associated with the release of endogenous pyrogens by rabbit or human leukocytes (131). As MDP is known to increase the release of prostaglandin E from macrophages (132), it was suggested that inhibition of the synthesis of these mediators of thermoregulation by indomethacin could inhibit the pyrogenic activity of MDP, and indeed indomethacin can inhibit the pyrogenicity of MDP without affecting its immunoenhancing properties (133), thus clearly dissociating pyrogenicity and adjuvant activity. It was later shown that MDP could directly interfere with thermosensitive brain structures, for a febrile response could be obtained after intracerebroventricular injection with a 10 000-fold smaller dose (134, 135). More recently it was found that several analogues are practically devoid of pyrogenicity although they are as adjuvant active as MDP (34).

5. Target cells and mechanism of action

In vivo experiments

The influence of MDP on the PFC (plaque forming cells) response to SRBC (sheep red blood cells) of mice has been studied. The effect of MDP depends on experimental conditions, it can even greatly suppress the antibody response, if injected at large doses and before immunization (79). It was furthermore shown that for the same doses of antigen and adjuvant injected simultaneously MDP inhibits the response when antigen is injected i.p. but enhances the response if antigen is injected i.v. (81). Löwy et al. (80), have shown the importance of T lymphocytes for the mediation of the adjuvant effect by using reconstituted irradiated mice; they first showed that T cells from mice sensitized in vivo to antigen plus MDP are able to mediate an increased response to SRBC; furthermore, the activity of spleen cells from animals sensitized 24 h earlier with SRBC and MDP was independent of the presence of either B cells or adherent cells but was completely abolished by treatment with anti θ-serum. These results show a mediation of antigen specific, MDP stimulated response by T cells, but they can not exclude a primary action of MDP on macrophages during the first 24 h of in vivo stimulation.

The importance of T cells for the mediation of MDP action in the restoration of the immune response has been demonstrated (136). It was known that removal of cells bearing specific surface receptors for an antigen, abrogated the immune response to this antigen, but recovery of the response can be accelerated by a process of unrelated delayed hypersensitivity. After removal of cells naturally able to give rosettes with pigeon erythrocytes and of cells leading to PFC against this antigen, the remaining spleen cells were transferred to an irradiated animal; the capacity to respond to an other antigen such as SRBC was maintained. Addition of MDP to pigeon erythrocytes led to an early specific recovery of T and B specific functions of the depleted population. This effect of MDP was T-cell dependent and furthermore a kinetic study of the restoration suggested that MDP could induce the proliferation of pluripotential B-cell precursors. The importance of T cells as a target for the mediation of the effect of MDP was also demonstrated by experiments showing that MDP can enhance carrier specific helper T cells in a carrier hapten system (43, 78, 137): for example (137) mice presensitized to keyhole limpet haemocyanine (KLH) with MDP, develop an increased anti TNP response upon immunization with TNP-KLH.

In vitro experiments

MDP can influence the *in vitro* immune response to SRBC; here again, MDP can inhibit or stimulate the response, according to the strain of mice and the cell density in cultures (138) and depending on the geometry of vessels used for the cultures and on the timing of addition of MDP (81); it is suggested that MDP can induce specific suppressive cells.

Most *in vitro* experiments were realized under conditions leading to an enhancement of the response by MDP. We have found that MDP can stimulate antigen sensitized lymphocytes, but the essential target seems to be an adherent cell. After a two hours contact with MDP, these cells are able to mediate an enhanced response in an adherent depleted spleen cell culture (Souvannavong & Adam, unpublished). It was shown (139) that the enhancement could be mediated by a supernatant from spleen cells incubated with MDP for 3 days. The release of the factor was unaffected by treatment with anti θ-serum but was completely inhibited by an antimacrophage serum; furthermore, active supernatants could be obtained from syngeneic or allogeneic adherent cells. Nevertheless, this active mediator released from macrophages by MDP does not allow a response in macrophage or T lymphocyte depleted spleen cells. These results suggest that MDP acts primarily on macrophages which release factors acting on B cells through T cell mediation. However, it was shown (140) that MDP can enhance the response in spleen cell cultures from nude mice; by using a microculture assay system (141) which permits an estimation of the frequency of precursor B cells, it has been shown (123) that MDP can replace T cells. It was suggested, that this activity is not due to a polyclonal activation, for it is dependent on the presence of antigen, but may result from a direct interaction of MDP with precursor B cells. But it was also shown in an *in vitro* study of the response to SRBC that B and T lymphocytes are the targets for the enhanced response induced by MDP (142); the

pretreatment of normal spleen cells for 4 to 24 h with MDP has no influence on PFC response to SRBC, but if spleen cells are first sensitized for 6 h with antigen, a 30 min contact with MDP is sufficient to induce an increased response. Optimal enhancement is nevertheless obtained when MDP is present during the whole period of culture. In a second set of experiments, antigen stimulated spleen cells have been separated into B or T lymphocytes and macrophages. The results show that B and T antigen sensitized cells are stimulated by MDP and are able to transfer an enhanced response.

MDP and macrophage functions

MDP and its derivatives like many microbial products (102) can affect most functions of the macrophage.

In vivo, as shown by Tanaka *et al.* (143), MDP increases the phagocytic activity of the reticulo-endothelial system, as measured by the rate of clearance of colloidal carbon. *In vitro*, it does not increase the capacity of resident mouse peritoneal macrophages to phagocytize SRBC (80) but it increases the phagocytosis and killing of *Listeria monocytogenes* (144) when acting on oil-induced guinea pig macrophages. This increased intracellular killing of *L. monocytogenes* can be tentatively linked to the ability of MDP to enhance the capacity of pharmacologically triggered mouse peritoneal macrophages to reduce oxygen to the superoxide anion (O_2^-) and then to hydrogen peroxide (145), which is probably involved in the killing of phagocytized microorganisms.

MDP is also able to increase the cytostatic, or cytolytic activity of mouse peritoneal macrophages. As first shown by Juy and Chedid (146), resident or thioglycollate elicited macrophages cultivated for 24 h in the presence of MDP, have an increased cytostatic activity against cells of the syngeneic P 815 mastocytoma. An increase of the cytostatic activity of resident macrophages can also be obtained by sequential *in vitro* treatment with TDM (cord factor) and MDP (147). MDP added *in vitro* to strongly cytostatic macrophages, obtained seven days after an i.p. injection of 50 μg of trehalose dimycolate (cord factor) in saline can prevent the loss of cytostatic activity due to *in vitro* cultivation (147).

A single injection of 30 μg of MDP i.p. four days

before harvest of the macrophages, does not increase their cytostatic activity, but three injections of doses of 1 mg or more, seven, five and three days before harvest (101) do so: the necessity of large and repeated doses might be due to the fast elimination of MDP (see below).

In a ^{51}Cr release assay, addition of MDP increases the cytolytic activity against tumour target cells of activated macrophages obtained from murine sarcoma virus induced tumours and from established murine macrophage cell lines (148). This effect is genetically regulated: for example a good response was obtained with Balb/c derived macrophages or cell lines but not with macrophages or cell lines from C_{57} Bl/6 mice; this correlates well with the results of Damais et al. (125), on the genetic dependance of the mitogenic response induced by MDP on splenocytes.

The mechanism of action of MDP in these experiments is still unknown. Although one cannot exclude that MDP could cause the release of a factor, such as interferon (148) which would activate the secreting macrophage itself or other macrophages in the same culture, a direct action of MDP on macrophages is the most probable alternative.

An example of a direct action is the inhibition of the migration of peritoneal macrophages from mice or guinea pigs (149, 150) induced by MDP. It was proved that the action of MDP is not linked to production of MIF but results from a direct interaction between MDP and adherent cells.

Various analogs of MDP have been assayed; a strict correlation was observed between the adjuvant activity of a compound and its ability to inhibit the migration of macrophages (149, 150).

MDP is also able to modify the spreading and adherence of macrophages (132, 144) and to increase their rate of incorporation of ^{14}C-glucosamine (151).

The macrophage is an actively secreting cell; MDP and its derivatives can modify the secretion pattern. Wahl et al. (132), have shown that secretion of the enzyme collagenase in vitro by oil-induced guinea pig macrophages was strongly enhanced by cultivation in the presence of MDP. As in the case of LPS, this increase is correlated with an increase in the secretion of PGE_2 which in turn influences the level of intracellular cAMP.

The synthesis and secretion of plasminogen activator by mouse peritoneal macrophages elicited by an i.p. injection of thioglycollate or killed Streptococci can be inhibited by cultivation in the presence of WSA (152) or MPP (N-acetylmuramyl-L-Ala-D-Glu-γ-meso-DAP-D-Ala-D-Ala) but not by MDP. These immunomodulators do not modify the concomitant secretion of lysozyme (71).

In some cases, MDP induces macrophages to secrete enhanced amounts of monokines, i.e. interleukines produced by cells of the monocyte-macrophage line: these monokines act on other cells which become the effector cells, the macrophage still being the target cell of MDP (139).

Wahl et al. (132) have shown that cultivation of oil-induced guinea pig peritoneal macrophages with MDP increases the production of a factor mitogenic for fibroblasts. MDP can also induce murine adherent peritoneal cells to produce a Colony Stimulating Factor (153). As shown by Oppenheim et al. (154), human monocytes cultivated with MDP secrete a factor mitogenic for thymocytes, which, according to its chromatographic pattern, is LAF (lymphocyte activating factor). Adherent peritoneal cells from BCG treated mice are also able to produce LAF upon stimulation with MDP; the same is true for the macrophage cell line P 388 D, but this production requires a long period of incubation which can be shortened by addition of activated spleen cells (154). The resident mouse peritoneal macrophages do not produce LAF and cannot be induced to produce it by MDP, but the background production of thioglycollate or trehalose dimycolate elicited macrophages can be increased several times by MDP and adjuvant active analogs (147); the half effect dose is about 0.2 μM, in the same range as the optimal concentration of MDP capable of inhibiting macrophage migration (150). In these experiments a relatively brief contact of MDP with the macrophages was enough to trigger the LAF production (70% of the effect after 1 hour of contact). This ability of MDP to induce or stimulate LAF secretion might be an important contribution to its immunopotentiating activities.

6. Metabolic fate in vivo of MDP and derivatives

When injected in saline MDP ^{14}C-, labelled on the C_1 of the lactyl group of muramic acid, is rapidly eliminated (70). 30 min after i.v. injection, more than

50% is recovered unchanged in the urine and 95% within 2 h. The uniformly ^{14}C labelled disaccharide pentapeptide GlcNAc-MurNAc-L-Ala-D-isoGln-γ-meso-DAP-D-Ala-D-Ala injected i.v. is also eliminated rapidly in the urine, partly as the intact molecule, partly as the corresponding pentapeptide (155): it thus seems that the disaccharide pentapeptide is recognized by an enzyme (an amidase hydrolyzing the linkage between MurNAc and L-Ala) which is inactive on MDP. Mur-NAc-L-Ala-D-Glu-γ-meso-DAP-^{14}C-D-Ala-^{14}C-D-Ala was also excreted rapidly in the urine, in part intact but mostly in the form of the tetrapeptide D-Glu-γ-meso-DAP-^{14}C-D-Ala-^{14}C-D-Ala, probably as a result of the sequential action of the amidase and an aminopeptidase splitting off the L-Ala residue. With Mur-NAc-L-Ala-D-Glu-γ-^{14}C-meso-DAP, the main product found in the urine is ^{14}C-meso-DAP, probably as the result of a degradation by a carboxypeptidase (Yapo, Parant, Petit, unpublished).

However, 45% of ^{14}C-MDP injected in a water in oil emulsion remain present at the injection site after 24 h (70). These results explain the necessity of much higher doses when MDP is injected in saline than when it is used in FIA. It also explains why repeated injections (101) or delivery through an osmotic minipump (119) produce results not obtained with a single injection.

Conclusions

The discovery of MDP has opened a new era in immunology, allowing for the first time a whole series of *in vivo* and, especially, *in vitro* studies of fundamental importance for immunology. But what about the future of MDP or its derivatives, as a clinically useful drug?

The last decade has witnessed the discovery of a great number of synthetic immunomodulators, such as levamisole, polynucleotides, lysolecithin analogues, etc. each of which has its characteristic activity pattern and some of which are already studied in large scale clinical trials.

MDP, the 'hero' of this review, has not yet undergone such trials, partly because of its pyrogenicity. It is, however, such an extraordinarily 'flexible' molecule, giving the inventive chemist thousands of possibilities of structural variations, that it is most probable, that the 'ideal' molecule(s) for clinical use will soon come forth. Moreover, we have seen, that depending on the structure of the three major moieties of MDP, one can produce molecules which are adjuvant, antibacterial and antitumour active, others, which are only antibacterial or antitumour active, etc. Pyrogenicity is no more a deterrent, as we have seen that the simple use of an antiinflammatory drug, such as indomethacin, eliminates fever, without diminishing adjuvancy; we also know that some simple derivatives have already been found, which are just as adjuvant and antibacterial as MDP, but practically devoid of pyrogenicity.

Let us thus foresee a clinical acceptable family of 'neo-MDP' drugs, which will be useful in the following situations:
– as adjuvants for veterinary and human vaccines, especially for viral vaccines with synthetic (but weakly immunogenic) vaccinating subunits,
– as anticancer and antiparasitic agents, either as vaccines, in combination with tumour or parasite antigens, or for stimulating non-specific resistance;
– as stimulants of non-specific resistance for treating infections with antibiotic resistant strains and for increasing the efficacy of antibiotics in immunodepressed patients, etc.

Let us, however, dampen the enthusiam of our readers, by reminding them of the potential hazards of the large-scale use of immunomodulators, as listed by Gisler *et al.* (156): (1) sensitization to immunostimulant itself, (2) sensitization to common extrinsic antigens, (3) sensitization to autoantigens, (4) increased formation of blocking antibodies, (5) increased formation of suppressor cells, (6) toxicity for myelolymphoid cells, (7) neoplastic transformation of myelolymphoid cells.

There is clearly still much work ahead in this field, for chemists, biologists and clinicians, but the benefit for Biology and Medicine will surely come.

Acknowledgements

The work of the authors was supported, by CNRS and, in part, by grants from DGRST, INSERM, Fondation de la Recherche Médicale, Ligue Nationale Française contre le Cancer, the Cancer Research Institute, New York, and a contract with Laboratoires Choay and Institut Pasteur.

44

Abbreviations

Ab: antibody; CMI: cell mediated immunity; DHS: delayed hypersensitivity; FCA: Freund's complete adjuvant; FIA: Freund's incomplete adjuvant; MDP: muramyldipeptide = N-acetyl-L-alanyl-D-isoglutamine; PFC: plaque forming cells; TDM: trehalose dimycolate = cord factor = P_3.

References

1. Freund, J., 1956. Advances in Tuberculogy, 7: 130–148.
2. Lederer, E., Adam, A., Ciorbaru, R., Petit, J. F. & Wietzerbin, J., 1975. Molecular & Cellular Biochem. 7: 87–104.
   Petit, J. F. & Lederer, E., 1978. Relations between structure and function in the prokaryotic cell. Symp. Soc. Gen. Biol. 28: 177–199.
3. Lederer, E., 1977. Med. Chem. V, pp. 257–279, Elsevier, Amsterdam.
4. White, R. G., Bernstock, L., Johns, R. G. & Lederer, E., 1958. Immunol. 1: 54–66.
   White, R. G., Jollès, P., Samour, D. & Lederer, E., 1964. Immunol. 7: 158–171.
5. Azuma, I., Kishimoto, Y., Yamamura, Y. & Petit, J. F., 1971. Japan J. Microbiol. 15: 193–197.
6. Adam, A., Ciorbaru, R., Petit, J. F. & Lederer, E., 1972. Proc. Natl. Acad. Sci. U.S.A. 69: 851–854.
7. Migliore-Samour, D. & Jollès, P., 1972. Biochem. Biophys. Res. Comm. 66: 1316–1321.
8. Hiu, I. J., 1972. Nature New Biology 238: 241–242.
9. Adam, A., Ciorbaru, R., Petit, J. F., Lederer, E., Chedid, L., Lamensans, A., Parant, F., Parant, M., Rosselet, J. P. & Berger, F. M., 1973. Infect. Immun. 7: 855–861.
10. Ciorbaru, R., Petit, J. F., Lederer, E., Zisssman, E., Bona, C. & Chedid, L., 1976. Infect. Immun. 13: 1084–1089.
11. Ciorbaru, R., Adam, A., Petit, J. F., Lederer, E., Bona, C. & Chedid, L., 1975. Infect. Immun. 11: 257–264.
12. Brochier, J., Bona, C., Ciorbaru, R., Revillard, J. P. & Chedid, L., 1976. J. Immunol. 117: 1434–1439.
13. Barot-Ciorbaru, R., Wietzerbin, J., Petit, J. F., Chedid, L., Falcoff, E. & Lederer, E., 1978. Infect. Immun. 19: 353–356.
14. Barot-Ciorbaru, R., Yakota, Y., Petit, J. F., Chedid, L. & Atanasiu, P., 1979. Ann. Microbiol. 130B: 263–269.
15. Stewart-Tull, D. E. S., Shimono, T., Kotani, S., Kato, M., Ogawa, Y., Yamamura, Y., Koga, T. & Pearson, C. M., 1975. Immunol. 29: 1–15.
16. Kumazawa, Y., Shibusawa, A., Suzuki, T. & Mizumoe, K., 1976. Japan J. Microbiol. 20: 183–190.
17. Nauciel, C., Fleck, J., Martin, J. P. & Mock, M., 1973. C.R. Acad. Sci. (série D) 276: 3499–3500.
18. Nauciel, C., Fleck, J., Martin, J. P., Mock, M. & Nguyên-Huy, H., 1974. Europ. J. Immunol. 4: 352–357.
19. Adam, A., Amar, C., Lederer, E., Petit, J. F. & Vilkas, E., 1974. C.R. Acad. Sci. (série D) 278: 799–801.
20. Adam, A., Ciorbaru, R., Ellouz, F., Petit, J. F. & Lederer, E., 1974. Biochem. Biophys. Res. Comm. 56: 561–567.
21. Kotani, S., Watanabe, Y., Shimono, T., Kinoshita, F., Narita, T., Kato, K., Stewart-Tull, D. E. S., Morisaki, I., Yokogawa, K. & Kawata, S., 1975. Biken J. 18: 93–104.
22. Nauciel, C., Fleck, J., Mock, M. & Martin, J. P., 1973. C.R. Acad. Sci. (série D) 277: 2841–2844.
23. Adam, A., Petit, J. F., Wietzerbin-Falszpan, J., Sinaÿ, P., Thomas, D. W. & Lederer, E., 1969. FEBS Letters 4: 87–92.
24. Azuma, I., Thomas, D. W., Adam, A., Ghuysen, J. M., Bonaly, R., Petit, J. F. & Lederer, E., 1970. Biochim. Biophys. Acta 208: 444–451.
25. Ellouz, F., Adam, A., Ciorbaru, R. & Lederer, E., 1974. Biochem. Biophys. Res. Comm. 59: 1317–1325.
26. Adam, A., Ellouz, F., Ciorbaru, R., Petit, J. F. & Lederer, E., 1975. Z. Immutätsforsch 149: 341–348.
27. Ellouz, F., 1975. Doctoral Dissertation, Fac. Pharm. Chatenay-Malabry.
28. Merser, C., Sinaÿ, P. & Adam, A., 1975. Biochem. Biophys. Res. Comm. 66, 1316–1322.
29. Kotani, S., Watanabe, Y., Kinoshita, F., Shimono, T., Morisaki, I., Shiba, T., Kusumoto, S., Tarumi, Y. & Ikenaka, K., 1975. Biken J. 18: 105–112.
30. Kotani, S., Narita, T., Stewart-Tull, D. E. S., Shimono, T., Watanabe, Y., Kato, K. & Iwata, S., 1975. Biken J. 18: 77–92.
31. Kotani, S., Watanabe, Y., Kinoshita, P., Schleifer, K. H. & Perkins, H. R., 1977. Biken J. 20: 1–4.
32. Chedid, L., Audibert, F. & Johnson, A. G., 1978. Prog. Allergy 25: 63–105.
33. Dukor, P., Tarcsay, L. & Baschang, G., 1979. Annu. Rep. Med. Chem. 14: 146–166.
34. Parant, M., 1979. Springer Semin. Immunopathol. 2: 101–118.
35. Lederer, E., 1980. J. Med. Chem. 23: 819–825.
36. Kusumoto, S., Tarumi, Y., Ikenaka, K. & Shiba, T., 1976. Bull. Chem. Soc. Japan 49: 544–539.
37. Lefrancier, P., Choay, J., Derrien, M. & Lederman, I., 1977. Int. J. Peptide Protein Res. 9: 249–257.
   Lefrancier, P., Derrien, M., Lederman, I., Nief, F., Choay, J. & Lederer, E., 1978. Int. J. Peptide Protein Res. 11: 289–296.
   Lefrancier, P., Petitou, M., Level, M., Derrien, M., Choay, J. & Lederer, E., 1979. Int. J. Peptide Protein Res. 14: 437–444.
38. Lefrancier, P. & Lederer, E., 1981. Progress Chem. Org. Nat. Prod. 40: 1–47. Springer, Wien.
39. Audibert, F., Parant, M., Damais, C., Lefrancier, P., Derrien, M., Choay, J. & Chedid, L., 1980. Biochem. Biophys. Res. Comm. 96: 915–923.
39a. Audibert, F., Parant, M., Chedid, L., Level, M., Lefrancier, P., Choay, J. & Lederer, E., 1981. Unpublished.
40. Chedid, L., Audibert, F., Lefrancier, P., Choay, J. & Lederer, E., 1976. Proc. Natl. Acad. Sci. USA 73: 2472–2475.
41. Audibert, F., Chedid, L., Lefrancier, P. & Choay, J., 1976. Cell. Immunol. 21: 243–249.
42. Azuma, I., Sugimura, K., Yamamura, Y., Kusumoto, S., Tarumi, Y. & Shiba, T., 1976. Jap. J. Microbiol. 20: 63–68.

43. Azuma, I., Sugimura, K., Taniyama, T., Yamawaki, M., Yamamura, Y., Kusumoto, S., Okada, S. & Shiba, T., 1976. Infect. Immun. 14: 18–27.

44. Adam, A., Devys, M., Souvannavong, V., Lefrancier, P., Choay, J. & Lederer, E., 1976. Biochem. Biophys. Res. Comm. 72: 339–346.

45. Hasegawa, A., Kaneda, Y., Amano, M., Kiso, M. & Azuma, I., 1978. J. Agricol. Biol. Chem. 42: 2187–2189.

46. Durette, P. L., Meitzner, E. P. & Shen, T. Y., 1979. Carbohyd. Res. 77: C1–C4.

47. Tsujimoto, M., Kinoshita, F., Okunaga, T., Kotani, S., Kusumoto, S., Yamamoto, K. & Shiba, T., 1979. Microbiol. Immunol. 23: 933–936.

48. Durette, P. L., Meitzner, E. P. & Shen, T. Y., 1979. Tetrahed. Lett. 4013–4016.

49. Nagai, Y., Akiyama, K., Kotani, S., Watanabe, Y., Shimono, T., Shiba, T. & Kusumoto, S., 1978. Cell. Immunol. 35: 168–172.

50. Parant, M., Damais, C., Audibert, F., Parant, E., Chedid, L., Sache, E., Lefrancier, P., Choay, J. & Lederer, E., 1978. J. Infect. Dis. 138: 378–386.

51. Chedid, L., Parant, M., Parant, F., Lefrancier, P., Choay, J. & Lederer, E., 1977. Proc. Natl. Acad. Sci. USA 74: 2089–2093.

52. Kiso, M., Kaneda, Y., Okumura, H., Hasegawa, A., Azuma, I. & Yamamura, Y., 1980. Carbohyd. Res. 79: C17–C19.

53. Hasegawa, A., Okumura, H., Kiso, M., Azuma, I. & Yamamura, Y., 1980. Carbohyd. Res. 79: C20–C23.

54. Kotani, S., Kinoshita, F., Morisaki, I., Shimono, T., Okunaga, T., Takada, H., Tsujimoto, M., Watanabe, Y., Kato, K., Shiba, T., Kusumoto, S. & Okada, S., 1977. Biken J. 20: 95–103.

55. Koga, T., Maeda, K., Onoue, K., Kato, K. & Kotani, S., 1979. Mol. Immunol. 16: 153–163.

56. Kotani, S., Watanabe, Y., Kinoshita, F., Morisaki, I., Kato, K., Shiba, T., Kusumoto, S., Tarumi, Y. & Ikenaka, K., 1977. Biken J. 20: 39–45.

57. Chedid, L., Parant, M. & Parant, F., 1977. Cr. Hebd. Séanc. Acad. Sci. Paris 284 D: 405–408.

58. Audibert, F., Chedid, L., Lefrancier, P., Choay, J. & Lederer, E., 1977. Ann. Immunol (Inst. Pasteur) 128 c.: 653–661.

59. Mašek, K., Zaoral, M., Jesek, J. & Krchnak, V., 1979. Experientia 35: 1397–1398.

60. Kotani, S., Kinoshita, F., Watanabe, Y., Morisaki, I., Shimono, T., Kato, K., Shiba, T., Ikenaka, K. & Tarumi, Y., 1977. Biken J. 20: 125–130.

61. Azuma, I., Sugimura, K., Yamawaki, M., Uemiya, M., Kusumoto, S., Okada, S., Shiba, T. & Yamamura, Y., 1978. Infect. Immun. 20: 600–607.

62. Yamamura, Y., Azuma, I., Sugimura, K., Yamawaki, M., Uemiya, M., Kusumoto, S., Okada, S. & Shiba, T., 1976. Gann. 67: 867–877.

63. Uemiya, M., Saiki, I., Kusoma, T., Azuma, I. & Yamamura, Y., 1979. Microbiol. Immunol. 23: 821–823.

64. Uemiya, M., Sugimura, K., Kusama, T., Saiki, I., Yamawaki, M., Azuma, I. & Yamamura, Y., 1979. Infect. Immun. 24: 83–89.

65. Azuma, I., Sugimura, K., Taniyama, T., Yamawaki, M., Yamamura, Y., Kusumoto, S., Okada, S. & Shiba, T., 1976. Infect. Immun. 14: 18–27.

66. Azuma, I., Yamawaki, M., Uemiya, M., Saiki, I., Panio, Y., Kobayashi, S., Fukuda, T., Imada, I. & Yamamura, Y., 1979. Gann. 76: 847–848.

66a. Takada, H., Tsujimoto, M., Kotani, S., Kusumoto, S., Inage, M., Shiba, T., Nagao, S., Yano, I., Kawata, S. & Yokogawa, K., 1979. Infect. Immun. 25: 645–652.

66b. Pabst, M. J., Cummings, N. P., Shiba, T., Kusumoto, S. & Kotani, S., 1980. Infect. Immun. 29: 617–622.

67. Parant, M., Audibert, F., Chedid, L., Level, M., Lefrancier, P., Choay, J. & Lederer, E., 1980. Infect. Immun. 27: 825–831.

68. Migliore-Samour, D., Bouchaudon, J., Floch, F., Zerial, A., Ninet, L., Werner, G. H. & Jollès, P., 1979. C.R. Acad. Sci. Paris, série D 289: 473–476.

69. Audibert, F., Heymer, B., Gros, C., Schleifer, K. H., Seidl, P. H. & Chedid, L., 1978. J. Immunol. 121: 1219–1222.

70. Parant, M., Parant, F., Chedid, L., Yapo, A., Petit, J. F. & Lederer, E., 1979. Int. J. Immunopharmacol. 1: 35–41.

71. Drapier, J. C., Lemaire, G., Tenu, J. P. & Petit, J. F., 1979. The Molecular Basis of Immune Cell Function (J. G. Kaplan, Ed.), Elsevier-North Holland, Biomedical Press, Amsterdam.

72. Reichert, C. M., Carelli, C., Jolivet, M., Audibert, F., Lefrancier, P. & Chedid, L., 1980. Mol. Immunol. 17: 357–363.

73. Chedid, L., Parant, M., Parant, F., Audibert, L., Lefrancier, P., Choay, J. & Sela, M., 1979. Proc. Natl. Acad. Sci. USA 76: 6557–6561.

74. Baschang, G., Dietrich, F. M., Gisler, R., Hartmann, A., Stanek, J. & Tarcsay, L., 1978. Swiss Patent 2035/78, Feb. 24.

75. Sela, M. & Mozes, E., 1979. Springer Semin. Immunopathol. 2: 119–132.

76. Wilkinson, P. C. & White, R. G., 1966. Immunology 11: 229–241.

76a. Souvannavong, V. T., Adam, A. & Lederer, E., 1978. Infect. Immun. 19: 966–971.

77. Tanaka, A., Saito, R., Sugiyama, K., Morisaki, I., Kotani, S., Kusumoto, S. & Shiba, T., 1977. Infect. Immun. 15: 332–334.

78. Löwy, I., Theze, J. & Chedid, L., 1980. J. Immunol. 124: 100–104.

79. Leclerc, C., Juy, D., Bourgeois, E. & Chedid, L., 1979. Cell. Immunol. 45: 199–206.

80. Löwy, I., Bona, C. & Chedid, L., 1977. Cell. Immunol. 29: 195–199.

81. Souvannavong, V. & Adam, A., 1980. Eur. J. Immunol. 10: 654–656.

82. Leclerc, C., Audibert, F. & Chedid, L., 1978. Immunology 35: 963–970.

83. Ohkuni, H., Norose, Y., Hayama, M., Kimura, Y., Kotani, S., Shiba, T., Kusumoto, S., Yokogawa, K. & Kawata, S., 1977. Biken J. 20: 131–136.

84. Ohkuni, H., Norose, Y., Ohta, M., Hayama, M., Kimura, Y., Tsujimoto, M., Kotani, S., Shiba, T., Kusomoto, S., Yokogawa, K. & Kawata, S., 1979. Infect. Immun. 24: 313–318.

85. Heymer, B., Finger, H. & Wirsing, C. H., 1978. Zeit. Immunität Forsch. 155: 87–92.
86. Kishimoto, T., Hirai, Y., Nakanishi, K., Azuma, I., Nagematsu, A. & Yamamura, Y., 1979. J. Immunol. 123: 2709–2715.
87. Audibert, F., Heymer, B., Gros, C., Schleifer, K. H., Seidl, P. H. & Chedid, L., 1978. J. Immunol. 121: 1219–1222.
88. Krause, R. M., 1975. Z. Immun. Forsch. Exp. Ther. 149S: 136–150.
89. Chedid, L., Carelli, C. & Audibert, F., 1979. J. Reticulo-endothelial. Soc. 26S: 631–641.
90. Chedid, L. & Audibert, F., 1976. In: P. A. Miescher, VIIth International Symposium of Immunopathology, Bad-Schache, pp. 382–396.
91. Toullet, F., Audibert, F., Voisin, G. A. & Chedid, L., 1977. Ann. Inst. Pasteur, Paris, 128C: 267–269.
92. Kozak, Y. de, Audibert, F., Thillaye, B., Chedid, L. & Faure, J. P., 1979. Ann. Immunol. (Institut Pasteur) 130C: 29–32.
93. Kohashi, O., Kotani, S., Shiba, T. & Ozawa, A., 1979. Infect. Immun. 26: 690–697.
93a. Nagao, S. & Tanaka, A., 1980. Infect. Immun. 28: 624–626.
94. Nagai, Y., Akiyama, K., Suzuki, K., Kotani, S., Watanabe, Y., Shimono, T., Shiba, T., Kusumoto, K., Ikuta, F. & Takeda, S., 1978. Cell. Immunol. 35: 158–167.
95. Nagai, Y., Akiyama, K., Kotani, S., Watanabe, Y., Shimono, T., Shiba, T. & Kusumoto, S., 1978. Cell. Immunol. 35: 168–172.
96. Audibert, F., Chedid, L. & Hannoun, C., 1977. C.R. Hebd. Séances Acad. Sci. Ser. D 285: 467–470.
97. Webster, R. G., Glezen, W. P., Hannoun, C. & Laver, W. G., 1977. J. Immunol. 119: 2073–2077.
97a. Okunaga, T., Kotani, S., Tada, Z., Kato, M., Kubo, T., Kusomoto, S., Inage, M. & Shiba, T., 1980. Nippon Saikingaku Zasshi Japan 35: 112.
98. Reese, R. T., Trager, W., Jensen, J. B., Miller, D. A. & Tantravahi, R., 1978. Proc. Natl. Acad. Sci. USA 75: 5665–5668.
98a. Mitchell, G. H., Richards, W. H. G., Voller, A., Dietrich, F. M. & Dukor, P., 1979. Bulletin World Health Organization 57: 189–197.
99. Siddiqui, W. A., Taylor, D. W., Kan, S. Ch., Kramer, K., Richmond-Crum, S. M., Kotani, S., Shiba, T. & Kusumoto, S., 1978. Science 201: 1237–1239.
100. Tribouley, J., Tribouley-Duret, J. & Appriou, M., 1979. C.R. Soc. Biol. 173: 1046–1049.
101. Matter, A., 1979. Cancer Immunol. Immunother. 6: 202–210.
102. Hibbs, J. B., Remington, J. S. & Stewart, C. C., 1980. Pharmac. Ther. 8: 37–69.
102a. Lederer, E., 1979. Springer Semin. Immunopathol. 2: 133–148.
103. Rapp, H. J., 1980. Isr. J. Med. Sci. 9: 366–374.
104. Zbar, B., Ribi, E., Meyer, T., Azuma, I. & Rapp, H. J., 1974. J. Natl. Cancer Inst. 52: 1571–1577.
105. Ribi, E., Milner, C., Granger, D. L., Kelly, M. T., Yamamoto, K., Brehmer, W., Parker, R., Smith, R. F. & Strain, S. M., 1976. Ann. New York Acad. Sci. 277: 228–238.
106. McLaughlin, C. A., Strain, S. M., Bickel, W. D., Goren, M. B., Azuma, I., Milner, K., Cantrell, J. L. & Ribi, E., 1978. Cancer Immunol. Immunother. 4: 61–68.
107. McLaughlin, C. A., Schwartzman, S. M., Horner, B. L., Jones, G. H., Moffat, J. G., Nestor, J. J. Jr. & Tegg, D., 1980. Science 208: 415–416.
108. Yarkoni, E., Lederer, E. & Rapp, H. J., 1981. Infect. Immun. 32: 273–276.
109. Yarkoni, E. & Rapp, H. J., 1980. Infect. Immun. 28: 881–886.
110. Yarkoni, E., Rapp, H. J., Polonsky, J. & Lederer, E., 1978. Int. J. Cancer 22: 564–569.
111. Parant, M., Parant, F., Chedid, L. & LeMinor, L., 1975. Ann. Immunol. (Paris) 126C: 321–326.
112. Parant, M., Parant, F. & Chedid, L., 1978. Proc. Natl. Acad. Sci. USA 75: 3395–3399.
113. Chedid, L., 1978. 7th Int. Congress Pharmacol. Paris, July.
114. Sackmann, W. & Dietrich, F. M., Abstract Int. Symp. on 'Infections in the Immunocompromised host', June 1980 Veldhoven (Eindhoven) The Netherlands.
115. Dietrich, F. M., Sackman, W., Zak, O. & Dukor, P., 1980. Curr. Chemother. Infect. Dis. Proc. Int. Congr. Chemother. 11th 1979. 2: 1730–1732.
116. Tobe, K., Nagata, Y. & Ogawa, H., 1980. Nippon Saikingaku Zasshi Japan 35: 175.
116a. Kujima, K., Tomioka, M., Nagao, S., Imat, K. & Tanaka, A., 1980. Nippon Saikingaku Zasshi Japan 35: 287.
117. Mathews, T. R. & Fraser-Smith, E. B., 1980. Curr. Chemother. Infect. Dis., Proc. Int. Congr. Chemother. 11th 1979, Eds. J. D. Nelson & C. Grassi, Am. Soc. Microbiol. 2: 1734–1735.
118. Finger, H. & Wirsing Von König, C. H., 1980. Infect. Immun. 27: 288–291.
119. Kierszenbaum, F. & Ferraresi, R. W., 1979. Infect. Immun. 25: 273–278.
120. Specter, S., Friedman, H. & Chedid, L., 1977. Proc. Soc. Exp. Biol. Med. 155: 349–352.
121. Damais, C., Parant, M., Chedid, L., Lefrancier, P. & Choay, J., 1978. Cell. Immunol. 35: 173–179.
122. Specter, S., Cimprich, R., Friedman, H. & Chedid, L., 1978. J. Immunol. 120: 487–491.
123. Watson, J. & Whitlock, C., 1978. J. Immunol. 121: 383–389.
124. Takada, H., Tsujimoto, M., Kotani, S., Kusumoto, S., Inage, M., Shida, T., Nagao, S., Yano, I., Kawata, S. & Yokogawa, K., 1979. Infect. Immun. 25: 645–652.
125. Damais, C., Parant, M. & Chedid, L., 1978. Antibiotics Chemother. 24: 19–29.
126. Möller, G., 1975. Transplant Rev. 23: 126–130.
127. Leclerc, C., Löwy, I. & Chedid, L., 1978. Cell. Immunol. 38: 286–293.
128. Löwy, I., Leclerc, C. & Chedid, L., 1980. Immunology 39: 441–450.
129. Löwy, I., Leclerc, C., Bourgeois, E. & Chedid, L., 1980. J. Immunol. 124: 320–325.
130. Kotani, S., Watanabe, Y., Shimono, T., Harada, K., Shiba, T., Kusumoto, S., Yokogawa, K. & Taniguchi, M., 1976. Biken J. 19: 9–13.
131. Dinarello, C. A., Elin, R. J., Chedid, L. & Wolff, S. M., 1978. J. Infect. Dis. 138: 760–767.

132. Wahl, S. M., Wahl, L. M., McCarthy, J. B. & Chedid, L., 1979. J. Immunol. 122: 2226–2231.

133. Parant, M., Riveau, G., Parant, F., Dinarello, C. A., Wolff, S. M. & Chedid, L., 1980. J. Infect. Dis. 124: 708–718.

134. Riveau, G., Masek, K., Parant, M. & Chedid, L., (1981) J. Exp. Med. 152: 869–877.

135. Masek, K., Kadlecova, O. & Petrovicky, P., 1980. Fever, Proc. Int. Symp. 1979 Ed. Lipton, J. M. Raven, New York, 123–130.

136. Prunet, J., Birrien, J. L., Panijel, J. & Liacopoulos, P., 1978. Cell. Immunol. 37: 151–161.

137. Sugimoto, M., Germain, R. N., Chedid, L. & Benacerraf, B., 1978. J. Immunol. 120: 980–982.

138. Leclerc, C., Juy, D. & Chedid, L., 1979. Cell. Immunol. 42: 336–343.

139. Fevrier, M., Birrien, J. L., Leclerc, C., Chedid, L. & Liacopoulos, P., 1978. Eur. J. Immunol. 8: 558–562.

140. Leclerc, C., Bourgeois, E. & Chedid, L., 1978. Immunol. Commun. 8: 55–64.

141. Lefkovits, I., 1972. Eur. J. Immunol. 2: 360–365.

142. Sugimura, K., Uemiya, M., Saiki, I., Azuma, I. & Yamamura, Y., 1979. Cell. Immunol. 43: 137–149.

143. Tanaka, A., Nagao, S., Saito, R., Kotani, S., Kusumoto, S. & Shela, T., 1977. Biochem. Biophys. Res. Comm. 77: 621–627.

144. Hadden, J. W., Englard, A., Sadlick, J. R. & Hadden, E. M., 1979. Int. J. Immunopharmac. 1: 17–27.

145. Pabst, M. J. & Johnston, R. B., 1980. J. Exp. Med. 151: 101–114.

146. Juy, C. & Chedid, L., 1975. Proc. Natl. Acad. Sci. USA 72: 4105–4109.

147. Tenu, J. P., Lederer, E. & Petit, J. F., 1980. Eur. J. Immunol. 10: 647–653.

148. Taniyama, T. & Holden, H. T., 1979. Cell. Immunol. 48: 369–374.

149. Yamamoto, Y., Nagao, S., Tanaka, A., Koga, T. & Onoue, K., 1978. Biochem. Biophys. Res. Comm. 80: 923–928.

150. Adam, A., Souvannavong, V. & Lederer, E., 1978. Biochem. Biophys. Res. Comm. 85: 684–690.

151. Takada, H., Tsujimoto, M., Kato, K., Kotani, S., Kusumoto, S., Inage, M., Shiba, T., Yono, I., Kawata, S. & Yokogawa, K., 1979. Infect. Immun. 25: 48–53.

152. Drapier, J. C., Lemaire, G. & Petit, J. F., 1980. In: Inflammation: Mechanisms and treatment (D. A. Willoughby, J. P. Giroud & N. Suartz, Eds), M.T.P. Press Ltd., Lancaster.

153. Staber, F. G., Gisler, R. H., Schumann, G., Tarcsay, L., Schäfli, E. & Dukor, P., 1978. Cell. Immunol. 37: 174–187.

154. Oppenheim, J. J., Tagawa, A., Chedid, L. & Mizel, S., 1980. Cell. Immunol. 50: 71–81.

155. Tomasic, J., Ladesic, B., Valinger, Z. & Hrsak, 1980. Biochim. Biophys. Acta 629: 77–82.

156. Gisler, R. H., Dietrich, F. M., Baschang, G., Brownbill, A., Schumann, G., Staber, F. G., Tarcsay, L., Wachsmuth, E. D. & Dukor, P., 1979. In: Drugs & Immune Responsiveness (J. L. Turk & Darien Parker, Eds), MacMillan Press, New York, pp. 133–160.

Received October 1, 1980.

Modulation of terminal deoxynucleotidyl transferase activity by thymosin

Shu-Kuang Hu*, Teresa L. K. Low and Allan L. Goldstein
Dept. of Biochemistry, The George Washington University School of Medicine and Health Sciences, Washington, DC 20037, U.S.A.
* *Present address: Sidney Farber Cancer Institute, Harvard Medical School Boston, MA 02115, U.S.A.*

Summary

Thymosin fraction 5, a family of acidic polypeptides isolated from bovine thymus, contains several hormonal-like factors which have been shown to influence the maturation, differentiation and functions of T-cells. Some of these peptides have been chemically defined. Two of them, thymosin α_1 (M.W. 3108) and thymosin β_4 (M.W. 4982) have been sequenced.

In murine systems, terminal deoxynucleotidyl transferase (TdT) has been shown to be T-cell specific and to be present primarily in the cortisone sensitive immature T-cell populations. The daily injection of thymosin fraction 5 and two of its components, thymosin β_3 and β_4, significantly increases TdT activity in immune suppressed mice as compared to control groups. This study indicates that thymosin can act on prothymocytes and influence the early stages of T-cell differentiation. In an in vitro system, thymosin fraction 5 and the purified peptide, thymosin α_1, have high activities in decreasing TdT in normal murine thymocytes after a 22-h incubation. This effect suggests that thymosin can also act on thymocytes and regulate the later biochemical processes during T-cell differentiation.

Introduction

It is now well established that a functioning thymus gland is an essential requirement of the normal development and maintenance of cell mediated immunity. Investigations on in vivo and in vitro effects of thymic extracts (1) and the effects of thymus grafts in cell-impermeable millipore diffusion chambers (2, 3, 4) have helped to establish the endocrine function of the thymus. In the past few years, factors possessing thymic hormone activity have been isolated from thymus tissue (5, 6, 7), blood (8) and thymic epithelial cell cultures. Most of the thymic hormones reported are polypeptides. One of the best characterized fractions is thymosin (1).

The thymosin studies were initiated at the Albert Einstein College of Medicine in 1964 by Abraham White and Allan L. Goldstein (9). Analytical iso-electric focusing of thymosin fraction 5, a partially purified preparation from calf thymus, has revealed the presence of a number of components (10). The molecular weights of these components are ranging from 1 000 to 15 000. A nomenclature based upon the isoelectric focusing pattern of thymosin fraction 5 in the pH range of 3.5–9.5 has been described and is illustrated in Fig. 1. Twenty of the thymosin peptides have now been purified to homogeneity. Two biologically active peptides termed thymosin α_1 and thymosin β_4 have been completely sequenced (10, 11). It appears that these thymosin peptides act individually, sequentially or in concert to influence the development of T-cell subpopulations. Clinically, bovine thymosin fraction 5 was reported to partially or fully reconstitute immune functions in humans with primary immunodeficiency diseases (12, 13) and cancer (14–16). Experimentally, thymosin fraction 5 and its purified components

Molecular and Cellular Biochemistry 41, 49–58 (1981).   0300-8177/81/0041–0049/$02.00.

50

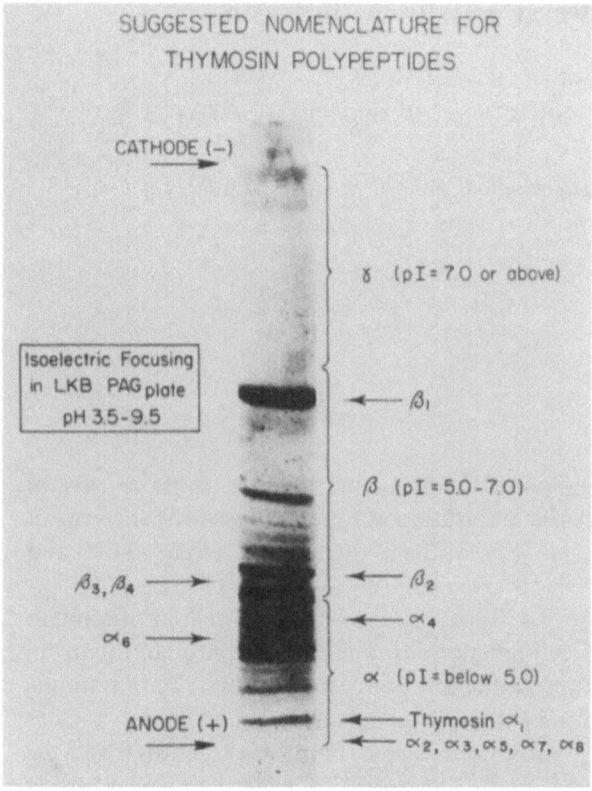

SUGGESTED NOMENCLATURE FOR
THYMOSIN POLYPEPTIDES

CATHODE (−)

δ (pI = 7.0 or above)

Isoelectric Focusing
in LKB PAG plate
pH 3.5-9.5

β₁

β (pI = 5.0-7.0)

β₃, β₄

β₂

α₆

α₄

α (pI = below 5.0)

ANODE (+)

Thymosin α₁

α₂, α₃, α₅, α₇, α₈

Fig. 1. Suggested nomenclature for thymosin polypeptides.

were shown to be very active in many in vivo and in vitro systems (17–23), including the induction of T-specific functions and markers.

Thymosin α_1 consists of 28 amino acid residues with a molecular weight of 3 108 and isoelectric point of 4.2 (10). The amino acid sequence of this peptide is shown in Fig. 2. The amino terminus of thymosin α_1 is blocked by an acetyl group. Chemically synthesized thymosin α_1 has been reported by Wang *et al.* (24) and shown as active as natural material in a few animal bioassays. Most recently, Wetzel *et al.* (25) have reported the isolation and complete chemical characterization of a N^α-desacetyl thymosin α_1 utilizing recombinant DNA procedures.

Thymosin β_4 consists of 43 amino acid residues with a molecular weight of 4 982 and isoelectric point of 5.1 (11). The amino acid sequence of this peptide is shown in Fig. 2. The N-terminal end of this peptide is also blocked by an acetyl group.

In the past several years, a number of enzymes have been isolated and found to be associated with specific lymphocyte populations (26, 27). One of these enzymes, terminal deoxynucleotidyl transferase (TdT), is T-cell specific and its level changes as T-cells undergo differentiation (28–30). During the

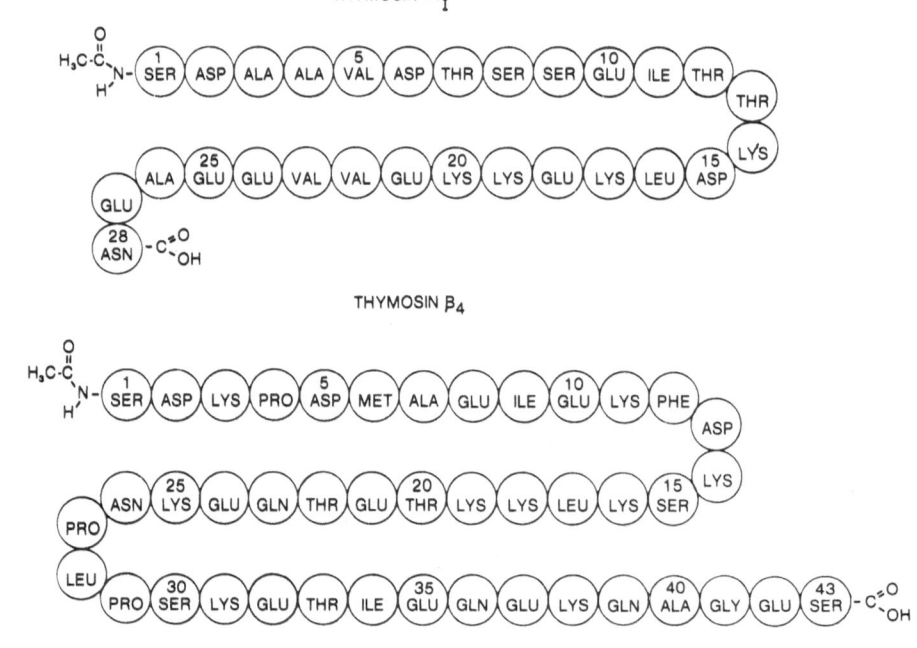

THYMOSIN α_1

H₃C-C-N-H— SER(1) ASP ALA ALA VAL(5) ASP THR SER SER GLU(10) ILE THR — THR — LYS — ASP(15) LEU LYS GLU LYS LYS(20) GLU VAL VAL GLU(25) GLU ALA — GLU — ASN(28) -C=O OH

THYMOSIN β_4

H₃C-C-N-H— SER(1) ASP LYS PRO ASP(5) MET ALA GLU ILE GLU(10) LYS PHE — ASP — LYS — SER(15) LYS LEU LYS LYS THR(20) GLU THR GLN GLU(25) LYS ASN — PRO — LEU — PRO SER(30) LYS GLU THR ILE GLU(35) GLN GLU LYS GLN ALA(40) GLY GLU SER(43) -C=O OH

Fig. 2. Amino acid sequence of thymosin α_1 and thymosyn β_4.

course of T-cell ontogeny, there is a progression from low TdT containing bone marrow stem cells to high TdT thymocytes, and then to TdT-negative peripheral T-cells. The exclusive localization of TdT in bone marrow stem T-cells and thymocytes suggests that TdT is associated with the early stages of T-cell differentiation. Thymosin fraction 5 (31, 32) and thymosin β_3 and β_4 (32) were found to induce TdT in bovine serum albumin gradient separated bone marrow cells from NIH Swiss nu/nu mice in vitro. In the studies described below, we also demonstrate that thymosin fraction 5 and thymosin β_3 and β_4 can greatly accelerate the reappearance of TdT positive cells in the thymus following steroid induced immunosuppression in an in vivo system. Thymosin fraction 5 and thymosin α_1 are highly active in depressing TdT activity in normal murine thymocytes in an in vitro system.

Materials and methods

Mice – Six-week old male C57Bl/6 mice were purchased from Jackson Laboratories, Bar Harbor, Maine. Four to six-week-old male BALB/c mice were obtained from Charles River Breeding Laboratories, Wilmington, Delaware.

TdT Preparation – Single thymocyte suspension from mouse thymus gland was prepared. No less than 5×10^7 cells were resuspended in CAK buffer (20 mM potassium cacodylate, 0.5 M KCl, 1 mM 2-ME, pH 7.5) to a final concentration of 10^8 cells/ml. Cells were pulse disrupted in an ice bath by a sonicator (Heat system Ultrasonics W-225 R, Plainview, N.Y.). Cell homogenate was ultracentrifuged at $100\,000 \times g$ for one hour. The supernatant was used as crude TdT extract.

TdT Assay – The TdT assay was adapted from Barton *et al.* (30). One unit of enzyme activity was defined as the amount catalyzing the incorporation of 1 nmole of dGTP into acid insoluble material/h, specific activity was calculated from the enzyme unit per 10^8 viable cells determined by trypan blue exclusion.

Thymosin Treatment in vivo – As diagrammed in Fig. 3, groups of six-week-old male C57Bl/6 mice with 6–8 animals/group were injected intraperitoneally with 1.25 mg hydrocortisone acetate (HCA) (Upjohn, Kalamazoo, MI)/mouse. After

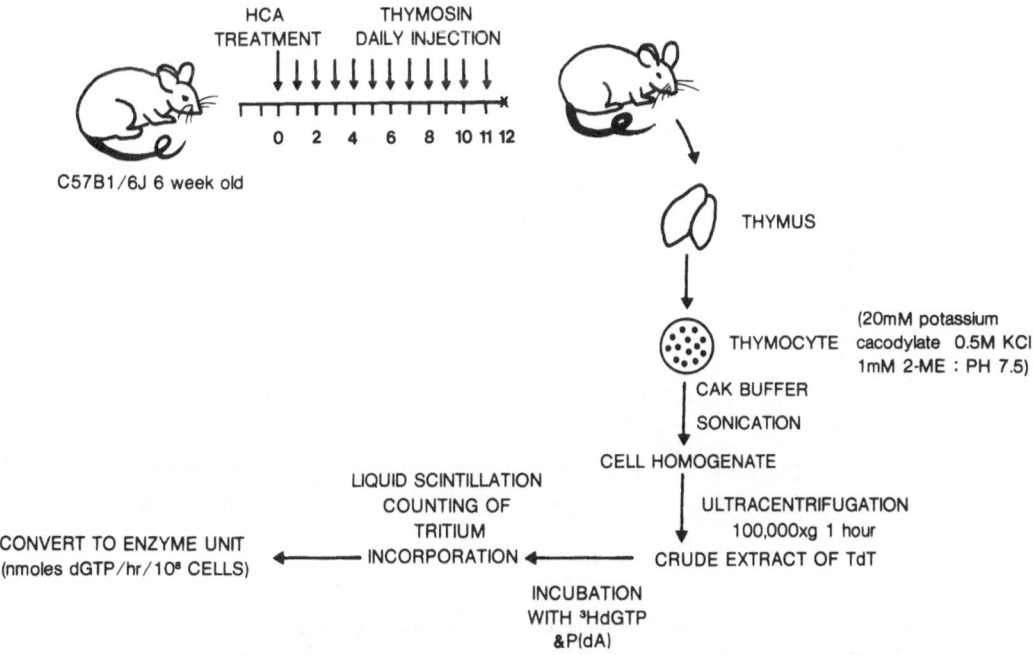

Fig. 3. In vivo TdT assay for thymosin. Six-week-old C57Bl/6J mice are treated with 1.25 mg HCA intraperitoneally followed by daily injection of thymosin for 11 days. Animals are sacrificed 24 h after the last injection. Thymocytes from individuals are pooled. TdT extract from thymocytes is obtained after sonication and ultracentrifugation. TdT specific activity is determined by enzyme assay.

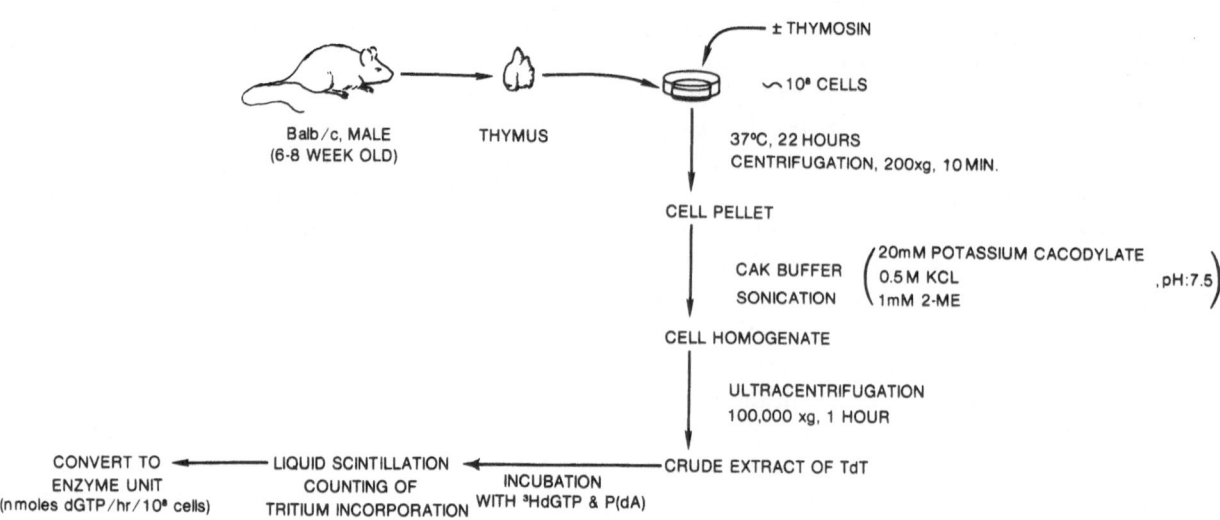

IN VITRO TdT ASSAY FOR THYMOSIN

Fig. 4. In vitro TdT assay for thymosin. 10^8 thymocytes from 6–8 week old BALB/c mice are incubated with thymosin fractions in HRPMI-1640 and 10% FCS for 22 h. Thymocytes are harvested. Crude extract of TdT is obtained by sonication following ultracentrifugation. TdT specific activity is determined by enzyme assay.

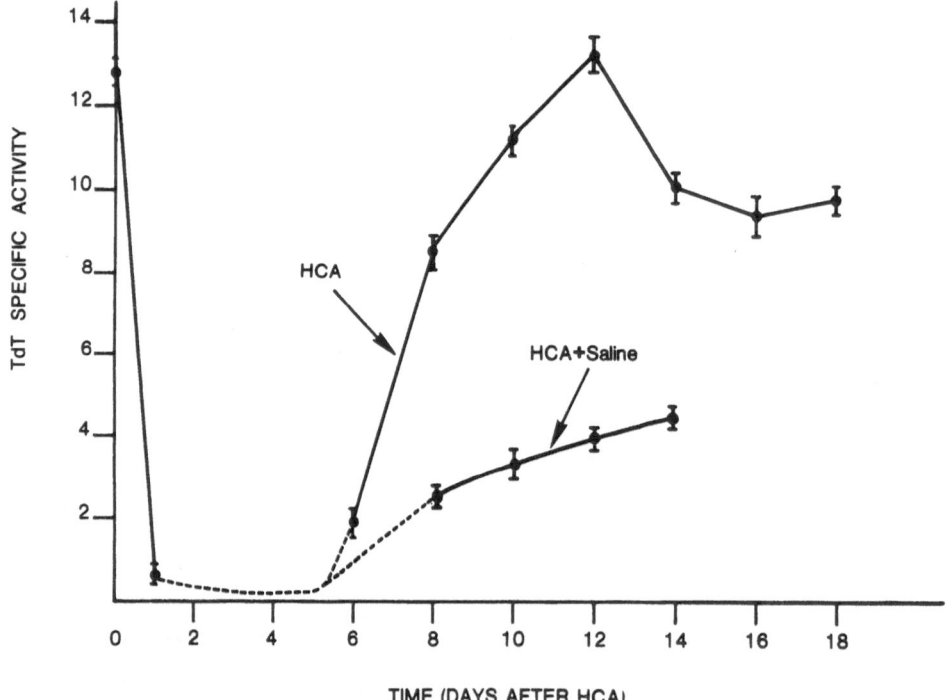

Fig. 5. Time course of TdT activity in involuted murine thymus caused by HCA or HCA followed by daily injection of saline. C57Bl/6J mice were treated with 1.25 mg HCA/animal intraperitoneally at day 0. The involution of TdT activity in the thymus started within 24 h after injection. TdT activity was undetectable from day 2 to day 4. A spontaneous recovery of TdT occurred around day 6. TdT level was the highest at day 12 and then back to the control level thereafter. The recovery of TdT activity in HCA treated thymus was depressed by daily injection of saline. Under the influences of HCA and daily injections, TdT recovery curve was lowered and went to a plateau after day 10. ●——● denotes enzyme activity which could not be detected by enzyme assay.

Table 1. Time-dependent effect of thymosin on TdT activity in HCA[a] treated C57B1/6J[b] thymus.

| Days after HCA treatment | TdT activity[c] | | % Increase |
	+ Saline	+ Thymosin fr. 5	
8	2.75 ± 0.27[d]	4.57 ± 0.23[d]	66.2
10	3.81 ± 0.34[d]	9.06 ± 0.48[d]	138.0
12	4.21 ± 0.21[d]	11.62 ± 0.51[d]	176.0
14	4.62 ± 0.23[d]	10.71 ± 0.37[d]	131.8

[a] 1.25 mg HCA/mouse injected peritoneally at day 0, followed by daily treatment of thymosin or saline.
[b] 6-week-old male animal.
[c] Expressed as nmoles of dGTP incorporated/10^8 cells/h.
[d] Mean ± S.E.

24 h, they received daily injections of thymosin fraction 5, spleen fraction 5 (prepared from calf spleen according to the same procedure as for thymosin fraction 5 (5)), purified thymosin peptides, control peptides or saline for 11 days. Mice were sacrificed 24 h after the last injection. Thymus glands were removed. TdT specific activity in thymocytes from each group was determined.

Thymosin Treatment in vitro – As diagrammed in Fig. 4, approximately 10^8 cells from 4–6 week old male BALB/c mouse thymus were placed in a sterile culture dish containing 20 ml HRPMI and 10% fetal calf serum. The cells were cultured with various thymosin peptides, control peptides or saline for 22 h at 37 °C in a 5% CO_2–95% air humidified incubator. Cells were harvested and TdT specific activity was determined.

Results

Effects of HCA and daily injection of saline on TdT activity in the thymus

The time-dependent recovery of TdT activity in murine thymus after a single injection of 1.25 mg HCA or 1.25 mg HCA followed by daily injection of saline for 11 days is shown in Fig. 5. These results indicate that a single injection of 1.25 mg HCA destroys TdT activity in the thymus but is followed by a spontaneous recovery of the enzyme activity. However, daily injection of saline after HCA treatment causes a hindrance to this spontaneous recovery.

Thymosin activity in increasing TdT activity

The ability of thymosin to increase TdT activity in involuted thymus glands following 1.25 mg HCA has been examined at four time points: Day 8, 10, 12 and 14. The results, shown in Table 1, indicate that daily injection of thymosin fraction 5 can significantly increase TdT activity in HCA treated thymus as compared to daily injection of saline. This effect is optimal at day 12. Spleen fraction 5, which was prepared using the same procedures developed for thymosin fraction 5, was tested in this system as a control fraction. Results in Table 2 indicate that thymosin fraction 5 is active in enhancing recovery of TdT activity after involution

Table 2. In vivo effect of thymosin on TdT activity in hydrocortisone acetate[a] teated C57B1/6J[b] thymocytes.

| Experiment | TdT specific activity[c] | | |
	+ Saline	+ Spleen fr. 5	+ Thymosin fr. 5
1	2.49	N.D.[e]	9.37 (3.76)
2	3.41	3.71 (1.09)[d]	10.90 (3.20)
3	2.61	2.56 (0.98)	4.39 (1.68)
4	4.21	4.68 (1.11)	8.72 (2.07)
Average ratio ± S.E.	1.00	1.06	2.68 ± 0.13 (p < 0.009)

[a] 1.25 mg HCA/mouse injected peritoneally at day 0, followed by daily treatment of thymosin or controls for 11 days.
[b] 6-week-old with 6–8 mice/group.
[c] Expressed as nmoles of dGTP incorporated/10^8 cells/hr.
[d] Ratio of TdT activity compared to saline group.
[e] N.D. = not determined.

54

Table 3. TdT activity in thymocytes from hydrocortisone acetate[a] treated C57B1/6J[b] mice following in vivo treatment with thymosin fractions.

Groups	TdT specific activity[c]	% Increase
Saline	4.91 ± 0.14[d]	
Spleen Fr. 5 (100 μg/day)	5.46 ± 0.22[d]	11
Thymosin Fr. 5 (100 μg/day)	10.62 ± 0.21[d]	116
Thymosin β_3 (1 μg/day)	11.06 ± 0.18[d]	125
Thymosin β_4 (1 μg/day)	10.03 ± 0.25[d]	104
Thymosin α_1 (1 μg/day)	6.20 ± 0.11[d]	26
Peptide β_1 (1 μg/day)	5.07 ± 0.13[d]	3

[a] 1.25 mg HCA/mouse injected peritoneally at day 0, followed by daily treatment of thymosin fractions or controls for 11 days.

[b] 6-week-old with 6–8 animals/group.

[c] Expressed as nmoles of dGTP incorporated/10^8 cells/h.

[d] Mean $\pm$ S.E.

caused by HCA and daily injections. Spleen fraction 5 was not active in this assay.

Since thymosin fraction 5 contains a family of peptides, some of its purified components were tested to identify the active factors for this specific effect. As shown in Table 3, the results indicate that thymosin β_3 and β_4 were active in inducing the reappearance of TdT. Thymosin α_1, which is very active in inducing other T cell functions, is not as active in this assay system. Peptide β_1, which is a predominant band on the isoelectric focusing gel of fraction 5 and found to be identical to ubiquitin and a portion of nuclear chromosomal protein A24 (13), did not show any activity.

The dose response of thymosin in this in vivo assay system (Fig. 6) suggests that daily injection of 10 μg of thymosin fraction 5 for 11 days causes a significant increase of TdT activity in HCA treated murine thymus glands. The effect is optimal when the concentration is increased to 100 μg/injection. The dose curve, however, maintains a plateau with increasing concentrations. The optimal dose for thymosin β_3 and β_4 is 1 μg/injection. The effects decrease at either higher or lower concentrations.

Thymosin activity in depressing TdT in normal thymocytes

In this assay system, whole thymocyte populations were used as target cells. As shown in Table 4, thymosin fraction 5 induced a significant decrease of TdT specific activity when compared to a medium control in thymocytes after a 22-h incuba-

tion, whereas spleen fraction 5 did not have any significant effect.

When further examining the activity of purified thymosin peptides in this assay system, the results shown in Table 5 demonstrate that only thymosin α_1 was able to induce the depression of TdT of all the thymosin peptides and controls tested. The time course study of this effect, shown in Table 6, indicates that incubation of 22 h gives the most optimum effect in depressing TdT.

The effect of thymosin α_1 in depressing TdT, as shown in Fig. 7, suggests that concentrations as low as 50 ng/ml provides a significant decrease of TdT activity in normal murine thymocytes after in vitro incubation. The optimal concentration of thymosin α_1 in depressing TdT was 150 ng/ml. However, the depression effect was diminished at higher doses. These results suggest that either thymosin α_1 specifically induces the decrease of TdT in mouse thymocytes within a range of concentration, or the overall effect at higher doses of thymosin α_1 was the net result of TdT depression and the TdT induction effect contributed from different cell populations.

Discussion

The rapid progress in the studies of thymosin experimentally and clinically has indicated that thymosin polypeptides regulate the maturation and functional development of thymic dependent lymphocytes. In the studies presented here, we have specifically focused on the influence of thymosin on

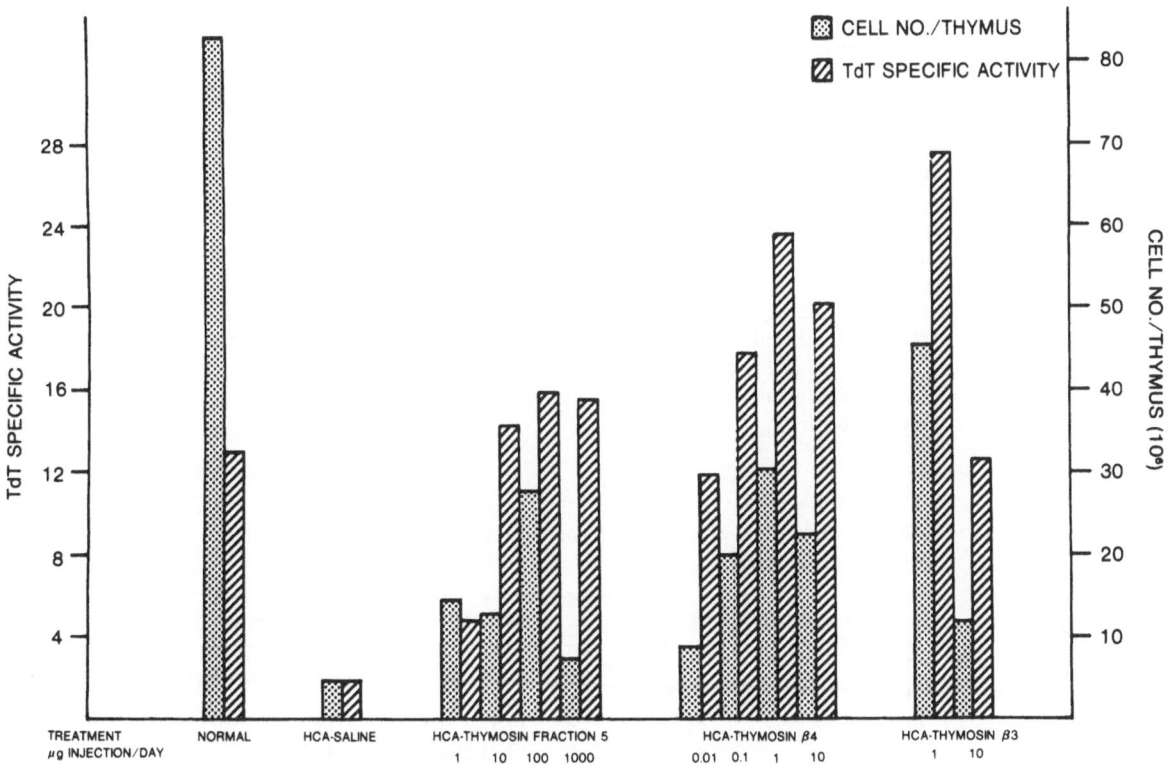

Fig. 6. Concentration dependence of the in vivo induction of TdT in HCA treated C57B1/6J thymus. Groups of 6-week-old C57B1/6J mice with 6-8 animals/group were treated with 1.25 mg HCA per animal intraperitoneally at day 0. After 24 h, they received daily injection of various doses of thymosin fraction 5, thymosin β_4, thymosin β_3 or saline for 11 days. TdT activity in pooled thymocytes from each group was measured 24 h after the last injection.

Table 4. In vitro effect of thymosin fraction 5 on depressing TdT activity in normal BALB/c thymocytes.

Experiments[a]	TdT specific activity[b]		
	+ Medium alone	+ Spleen fr. 5	+ Thymosin fr. 5
1	12.5	10.9 (0.87)[c]	7.1 (0.57)
2	10.7	10.1 (0.94)	6.8 (0.64)
3	13.1	12.1 (0.92)	8.9 (0.68)
4	10.5	9.7 (0.92)	7.2 (0.69)
Average ratio[c] ± S.E.	1.00	0.91 ± 0.04	0.65 ± 0.08
	$p = 0.28$[d]		$p = 0.002$[d]
		$p = 0.004$[d]	

[a] 10^8 BALB/c thymocytes were incubated ± thymosin fraction 5 (500 ng/ml) or spleen fraction 5 (500 ng/ml) for 22 h at 37 °C in 20 ml of HRPMI + 10% FCS.

[b] Expressed as nmoles of dGTP incorporation/10^8 cells/h.

[c] Ratio of TdT activity compared to medium control group.

[d] One way analysis of variance.

Table 5. In vitro effect of thymosin peptides on depressing TdT activity in normal BALB/c thymocytes.

Treatment[a]	Concentration (ng/ml)	TdT activity[b]	% Decrease
Medium alone		12.1 ± 0.2[c]	
BSA	500	11.9 ± 0.3[c]	1.7
Peptide β_1	150	11.4 ± 0.2[c]	5.8
Thymosin β_3	100	12.3 ± 0.2[c]	–
Thymosin α_1	50	6.8 ± 0.3[c]	43.8
Thymosin Fr. 5	500	7.3 ± 0.2[c]	39.7

[a] 10^8 BALB/c thymocytes were incubated with thymosin peptides or controls for 22 h @ 37 °C in 20 ml of HRPMI + 10% FCS.

[b] Expressed as nmoles of dGTP incorporation/10^8 cells/h.

[c] Mean ± S.E.

Table 6. Time effect of thymosin α_1 on depressing TdT activity in normal BALB/c thymocytes.[c]

Incubation period (h)	% Viability[b]	TdT activity[a]	% Decrease
0	99.3	12.31 ± 0.48[d]	
4	98.7	12.47 ± 0.40[d]	–
10	94.0	10.92 ± 0.38[d]	11.3
22	81.3	7.31 ± 0.35[d]	40.6
36	51.2	7.47 ± 0.41[d]	39.3

[a] Expressed as nmoles of dGTP incorporation/10^8 cells/h.

[b] Determined by trypan exclusion.

[c] 10^8 thymocytes from 6-week-old male BALB/c mice were incubated with 150 ng thymosin α_1 for various time period at 37 °C in 20 ml of HRPMI + 10% FCS.

[d] Mean ± S.E.

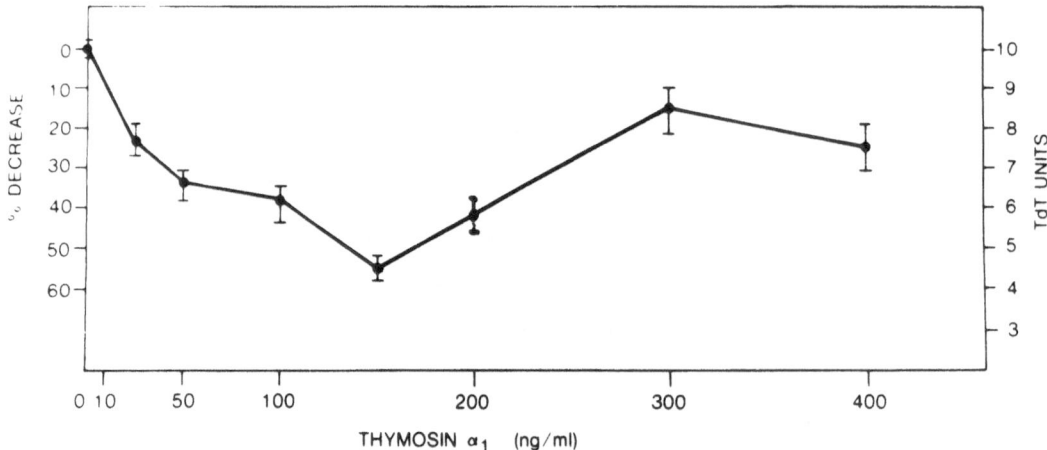

Fig. 7. Dose response of thymosin α_1 in depressing TdT activity in normal murine thymocytes. Approximately 10^8 thymocytes from 6–8 week old BALB/c mice were incubated with various concentrations of thymosin α_1 or controls for 22 h at 37 °C in HRPMI containing 10% FCS. TdT specific activity was measured by enzyme assay. Thymosin α_1 induced depression of TdT activity in thymocytes was expressed as percent decrease.

the early stages of thymocytes differentiation. Although the in vivo role of TdT is still a mystery, its unique distribution in immature T-cells suggests that TdT is a good marker for immature T cells (26, 28, 29, 30).

In our in vivo TdT assay system, HCA was used to destroy the TdT(+) thymocytes. According to the spontaneous recovery of TdT activity (Fig. 5), a high TdT-containing cell population was present in the thymus at day 12 after HCA injection. However, this time-dependent curve changes if HCA treatment is followed by daily injections of saline. In this case, the recovery of TdT activity in the thymus is highly depressed. It might be caused by the continuous release of steroid in the animal induced by the injections. The results of the in vivo TdT assay indicated that thymosin fraction 5 and two of its component peptides, thymosin β_3 and β_4, significantly increase TdT activity in HCA treated thymus and overcomes the depression of TdT recovery as seen in the HCA-saline group.

The results of the in vitro TdT model indicate that thymosin fraction 5 and its purified component thymosin α_1 decrease TdT levels in normal murine thymocytes after incubation as compared to control groups. Since the in vivo differentiation of

thymocytes occurs under the well-toned intrathymic microenvironment, there might be a series of differential steps involved in the whole process. Therefore, it is reasonable that thymosin decreases TdT in the immature cells only to the level expressed in more mature thymocytes instead of totally turning off the enzyme activity under the in vitro condition.

Most recently, using an indirect fluorescent antibody technique, Hirokawa et al. (personal communication) demonstrated that thymosin β_3 was detected specifically in epithelial cells covering the cortical surface, whereas thymosin α_1 was detected specifically in thymic epithelial cells of the medulla. His finding and the distribution of TdT in the normal thymocyte populations strongly support our TdT data, indicating the activity of thymosin in inducing T cell differentiation. That is, prothymocytes migrate to the thymic cortex and become high TdT-containing cortical thymocytes under the influence of thymosin β_3. Some of these thymocytes then move to the thymic medulla and become TdT(-), more mature thymocytes under the influence of thymosin α_1.

Based upon the studies of thymosin in modulating TdT activity and the results of previous findings, a proposed role of thymosin peptides in T cell

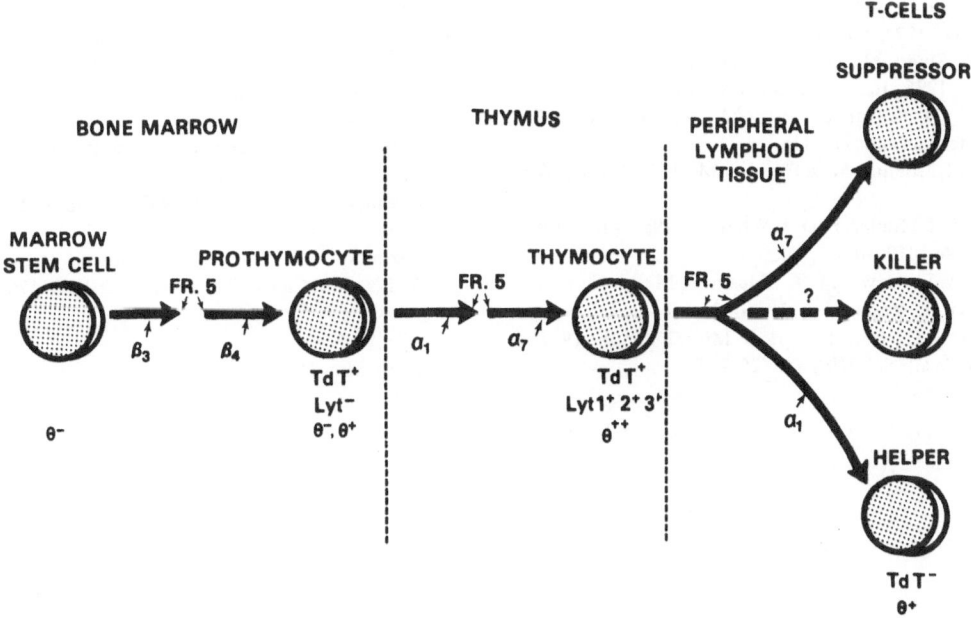

Fig. 8. Proposed role of thymosin peptides in T-cell maturation. Based upon the studies of thymosin in modulation of T-specific markers and functions, an overall scheme of roles of thymosin peptides in T-cell differentiation is proposed.

58

differentiation can be postulated, as shown in Fig. 8.

Expanding our understanding of the biological roles of individual thymosin polypeptides will be of great importance in establishing the basic mechanisms whereby lymphocytes mature and differentiate, as well as in utilization of thymosin in the treatment of diseases that manifest various immunological abnormalities.

Acknowledgements

Research for this paper was supported in part by grants from the National Cancer Institute, CA 29943 and CA 24974 and Hoffmann-La Roche, Inc.

References

1. Low, T. L. K. & Goldstein, A. L., 1978. In: 'The Year in Hematology' (Silber, R., LoBue, J. & Gordon, A. S., eds), Plenum Press, New York, pp. 281–319.
2. Levey, R. H., Trainin, N. & Law, L. W., 1963. J. Natl. Cancer Inst. 31: 199–217.
3. Osoba, D. & Miller, J. F. A. P., 1964. J. Exp. Med. 119: 177–194.
4. Dougherty, T. F., Berliner, M. L., Schneebeli, G. L. & Berliner, D. L., 1964. Ann. N.Y. Acad. Sci. 113: 825–843.
5. Hooper, J. A., McDaniel, M. C., Thurman, G. B., Cohen, G. H., Schulof, R. S. & Goldstein, A. L., 1975. Annals N.Y. Acad. Sci. 249: 125–144.
6. Goldstein, G., 1974. Nature 247: 11–14.
7. Trainin, N., Small, M., Zipori, D., Umiel, T., Kook, A. I. & Rotter, V., 1975. In: 'The Biological Activity of Thymic Hormones' (D. K. van Bekkum, ed.) Kooyker Scientific Publ., Rotterdam, pp. 117–114.
8. Bach, J. F., Dardenne, M. & Pleau, J. M., 1977. Nature 266: 55–57.
9. Goldstein, A. L., Slater, F. D. & White, A., 1966. Proc. Natl. Acad. Sci. 56: 1010–1017.
10. Goldstein, A. L., Low, T. L. K., McAdoo, M., McClure, J., Thurman, G. B., Rossio, J. I., Lai, C.-Y., Chang, D., Wang, S.-S., Harvey, C., Ramel, A. H. & Meienhofer, J., 1977a. Proc. Natl. Acad. Sci. USA 74: 725–729.
11. Low, T. L. K., Hu, S.-K. & Goldstein, A. L., 1981. Proc. Natl. Acad. Sci. USA 78: 1162–1166.
12. Wara, D. W., Goldstein, A. L., Doyle, W. & Ammann, A. J., 1975. N. Eng. J. Med. 292: 70–74.
13. Wara, D. W. & Ammann, A. J., 1976. Pediatr. 57: 643–646.
14. Goldstein, A. L., Cohen, G. H., Rossio, J. L., Thurman, G. B., Ulrich, J. T. & Brown, C. N., 1976. Med. Clin. N. Am. 60: 591–606.
15. Schafer, L. A., Goldstein, A. L., Gutterman, J. U. & Hersh, E. M., 1976. Ann. N.Y. Acad. Sci. 277: 607–620.
16. Schafer, L. A., Gutterman, J. U., Hersh, E. M., Mavligit, G. M. & Goldstein, A. L., 1977. In: 'Progress in Cancer Research and Therapy (Chirigos, M. A., ed.) Vol. 2, pp. 329–346, Raven Press, New York.
17. Thurman, G. B., Ahmed, A., Strong, D. M., Gershwin, M. E., Steinberg, A. D. & Goldstein, A. L., 1975. Transplant. Proc. 7: 299–303.
18. Bach, J. F., Dardenne, M., Goldstein, A. L., Guha, A. & White, A., 1971. Proc. Natl. Acad. Sci. USA 68: 2734–2738.
19. Dauphinee, M. J., Talal, N., Goldstein, A. L. & White, A., 1974. Proc. Natl. Acad. Sci. USA 71, 2637–2641.
20. Zisblatt, M., Goldstein, A. L., Lilly, F. & White, A., 1970. Proc. Natl. Acad. Sci. USA 66: 1170–1174.
21. Hardy, M. A., Zisblatt, M., Levine, N., Goldstein, A. L., Lilly, F. & White, A., 1971. Transplant. Proc. III: 926–928.
22. Khaw, B. A. & Rule, A. H., 1973. Br. J. Cancer 28: 288.
23. Scheinberg, M. A., Goldstein, A. L. & Cathcart, E. S., 1976. J. Immunol. 116: 156–158.
24. Wang, S.-S., Kulesha, I. D. & Winter, D. P., 1978. J. Am. Chem. Soc. 101: 253–254.
25. Wetzel, R., Heyneker, H. L., Goeddel, D. V., Jhurani, P., Shapiro, J., Crea, R., Low, T. L. K., McClure, J. E. & Goldstein, A. L., 1980. Biochemistry 19: 6096–6104.
26. Chang, L. M. S., 1971. Biochem. Biophys. Res. Commun. 44: 124–131.
27. Weinstein, Y., 1977. J. Immunol. 119: 1223–1229.
28. Coleman, M. S., Hutton, J. J. & Bollum, F. J., 1974. Biochem. Biophys. Res. Commun. 58: 1104–1109.
29. Gregoire, I. E., Goldschneider, I., Barton, R. W. & Bollum, F. J., 1977. Proc. Natl. Acad. Sci. 74: 3993–3996.
30. Barton, R., Goldschneider, I. & Bollum, F. J., 1976. J. Immunol. 116: 462–468.
31. Pazmino, N. H., Ihle, J. N. & Goldstein, A. L., 1978. J. Exp. Med. 147: 708–718.
32. Pazmino, N. H., Ihle, J. N., McEwan, R. N. & Goldstein, A. L., 1978. Cancer Treat. Rep. 62: 1749–1755.

Received May 15, 1981.

Isolation of three separate anaphylatoxins from complement-activated human serum

Tony E. Hugli, Craig Gerard, Marleen Kawahara, Maurice E. Scheetz II, Russel Barton, Stephen Briggs, Gary Koppel and Susan Russell

Molecular Immunology Scripps Clinic & Research Foundation La Jolla, CA 92037 U.S.A., Lilly/Scripps Scripps Clinic & Research Foundation La Jolla, CA 92037 U.S.A., Lilly Research Laboratories Division of Eli Lilly and Co. Indianapolis, IN 46204 U.S.A.

Summary

Recent methodologies used in preparing anaphylatoxins from complement-activated serum are described. Activation of the alternative pathway generates C3a and C5a; however, activation of the classical pathway is required to generate the anaphylatoxin from C4. This article describes an activation scheme that simultaneously generates all three of the anaphylatoxins (e.g., C3a, C4a and C5a) in human serum and outlines a procedure for isolating each as homogeneous products. Purification of intact anaphylatoxins directly from complement-activated serum takes place only if an exopeptidase in serum, known as carboxypeptidase N (SCPN), is properly inhibited. A new series of mercapto derivatives of arginine analogs are introduced as potent and effective inhibitors of SCPN. These inhibitors permit normal complement activation but prevent degradation of the released activation fragments C3a, C4a or C5a.

The SCPN inhibitor previously used was 6-aminohexanoic acid (EACA), but it required a 1 M concentration for effective inhibition, the substituted mercapto-guanido compounds prove to be effective in the mM range.

Introduction

Research concerning the anaphylatoxins has been moving very rapidly (1); however, up to this point one major aspect of the field has been largely overlooked. It is terribly important to take stock of the recent progress and to have a detailed description of the most current methodology that is used in the isolation of anaphylatoxins. A survey of the new procedures employed in obtaining fully active products from complement-activated serum is presented.

In the past several years a third anaphylatoxin, C4a, (2, 3) has been described and this factor joins the ranks of C3a and C5a as potent humoral mediators possessing numerous biological activities (4, 5). Characterization of the biological attributes of these complement by-products continues to reveal new and exciting functions. Activities that

are expressed by the anaphylatoxins imply a major role in host defense mechanisms such as inflammation and in more specialized mechanisms such as hypersensitivity. In general, it is true that the biological activity of C5a exceeds that of C3a which in turn is more potent than C4a. Furthermore, the biological diversity of C5a exceeds that of C3a and C4a; although each factor exhibits spasmogen activity, only C5a is a potent chemotactic factor (6). Recent findings have expanded further the known functional diversity of the C5a molecule by demonstrating an ability of this anaphylatoxin to stimulate synthesis and release of arachidonate products known as leukotrienes from pulmonary tissue (7, 8). This observation greatly enhances the recognized potential of anaphylatoxins in vivo by providing evidence that these humoral mediators induce the release of a variety of cellular mediators from granulocytes that are considerably more complex

Molecular and Cellular Biochemistry 41, 59–66 (1981).   0300-8177/81/0041-0059/$01.60.

than the vasoamines.

Other recent reports have described degradation products of C5a that prove to be chemotactic for tumor cells, but not for cell types such as normal leukocytes that are responsive to the intact factor (9, 10). This new observation has wide implications for protective surveilance mechanisms in the host's immune defense system. The true role of anaphylatoxin in this mechanism will be understood only after the degradation products have been isolated and further characterized. This, of course, brings us back to the original theme which is to examine current methodology used in isolating the anaphylatoxins directly from activated human serum and to consider the importance of having valid chemical characterizations before biologic studies are undertaken.

Materials and methods

The chemicals anisil alcohol, dimethylaniline, trichloroacetylchloride, p-methoxybenzyloxy-carbonyl azide ethylnipecotate and thiolacetic acid were obtained from Aldrich Chem. Co., MCB and Mallinckrodt. Other chemicals used in the preparation of (dl) 2-mercaptomethyl-5-guanidinopentanoic acid were of reagent grade. The ammonium formate used in the chromatographic separation of the anaphylatoxins was obtained from Aldrich Chem. Co. SP-Sephadex and QAE-Sephadex were from Pharmacia and the CM-Cellulose was purchased from Whatman as CM-52. The Bio gel P-60 was obtained from Bio Rad.

The synthesis of (dl) 2-mercaptomethyl-5-guanidinopentanoic acid.

STEP 1.

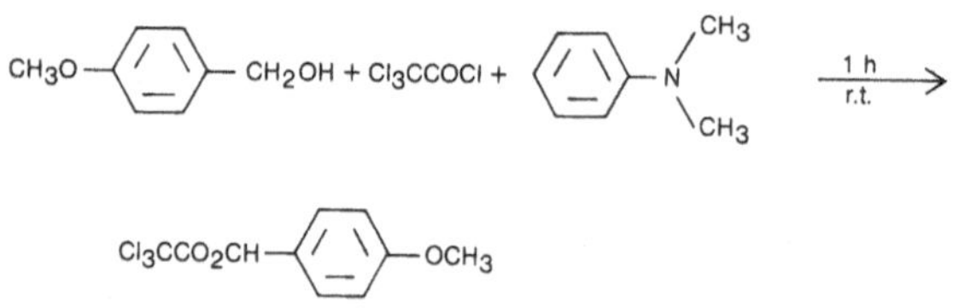

(i)

Procedure: 760 g (5.5. M) of anisil alcohol and 670 g (5.5. M) of dimethylaniline were dissolved in 5 l of toluene under a nitrogen atmosphere and cooled to 0 °C. 1 kg (5.5. M) of trichloroacetylchloride in 1 l of toluene was then added slowly. The reaction was allowed to reach room temperature and was stirred for 1 h. This mixture was poured into ice, then washed with 10% sulfuric acid, water, NaHCO$_3$, and again with water. The final product was dried over Na$_2$SO$_4$ and concentrated. Yield = 1,600 g of p-methoxy-benzyltrichloroacetate (I).

STEP 2.

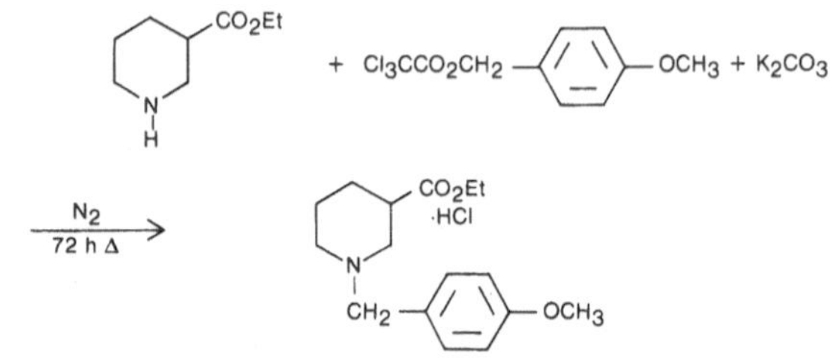

(II)

Procedure: 695 g (4.4 M) of ethylnipecotate, 1,304 g (4.8 M) of p-methoxy-benzyltrichloroacetate and 660 g (4.8 M) of potassium carbonate in 7 l of toluene were refluxed for 72 h under a nitrogen atmosphere. The reaction mixture was cooled and solvent was removed under reduced pressure. The resulting oil was taken up in chloroform, washed with 10% potassium carbonate and 10% hydrochloric acid, then dried over Na$_2$SO$_4$, and finally concentrated under reduced pressure. The resulting oil was triturated with ether, yielding 783 g of solid ethyl N-(p-methoxybenzyl) nipecotate hydrochloride (II).

STEP 3.

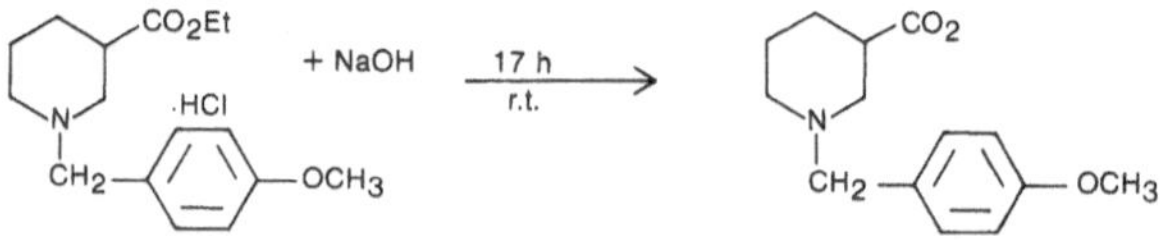

(III)

Procedure: 783 g (2.5 M) of ethyl N-(p-methoxybenzyl) nipecotate hydrochloride, 211 g (5.3 M) of sodium hydroxide, 1 l of water and 5 l of methanol were stirred at room temperature for 17 h. Solvents were removed under reduced pressure, toluene was added and later removed under reduced pressure.

Crude yield 1,055 g of N-(p-methoxybenzyl) nipecotic acid (III).

STEP 4.

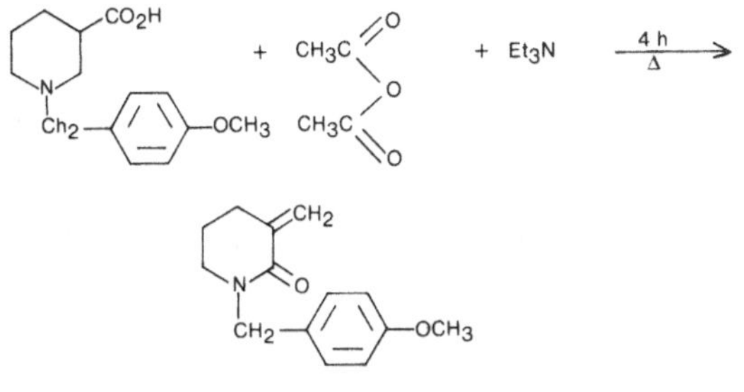

(IV)

Procedure: 800 g (crude) of N-(p-methoxybenzyl) nipecotic acid, 6 l of acetic anhydride and 1 l of triethylamine were refluxed for 4 h. Solvents were removed under reduced pressure. The product was taken up in chloroform, washed with water, dried over sodium sulfate and concentrated. The resulting oil was chromatographed over silica gel, using hexane-ethylacetate (1:1) as eluent. Yield 219 g (Rf 0.4) of 1-(p-methoxybenzyl)-3-methylene-2-piperidone (IV).

STEP 5.

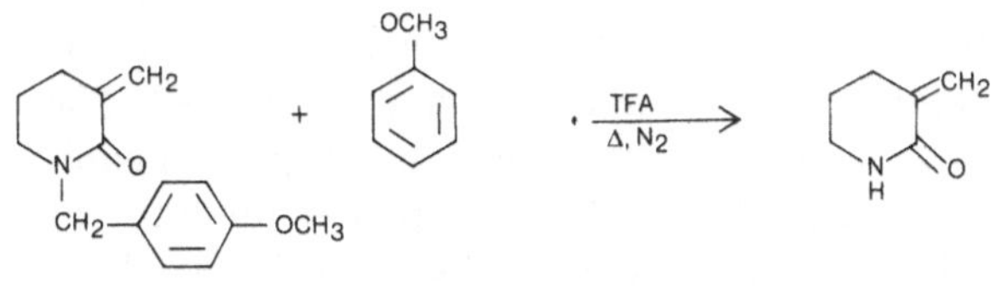

(V)

Procedure: 219 g (0.99 M) of 1-(p-methoxybenzyl)-3-methylene-2-piperidone, 276 g (2.6 M) of anisole and 3 l of TFA were refluxed for 48 h under a nitrogen atmosphere. Solvents were removed under reduced pressure. The residue was chromatographed over silica gel, using ethylacetate as eluent. Yield 115 g (Rf 0.2) of 3-methylene-2-piperidone (V).

STEP 6.

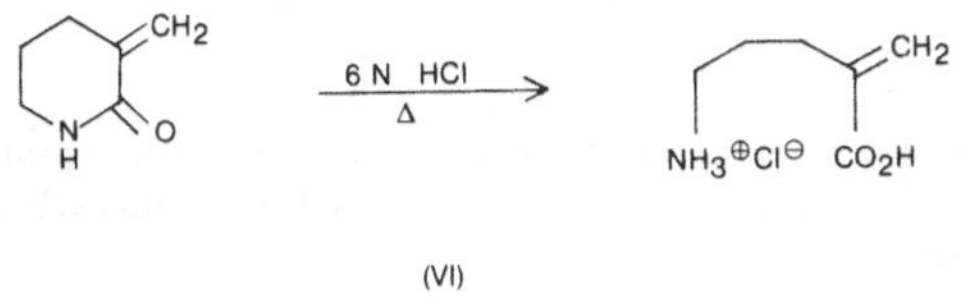

(VI)

Procedure: 177 g (1.59 M) of 3-methylene-2-piperidone and 8 l of 6 N hydrochloric acid were refluxed for 40 h. The mixture was cooled and extracted with methylene chloride, the aqueous layer was concentrated under reduced pressure, toluene was added, and solvents were again removed under reduced pressure. The residue was recrystallized from isopropanol and after extracting the product with hot isopropanol. Yield 44.6 g (solid); second crop yield 26.9 g (oil) of 2-methylene-5-aminopentanoic acid HCl (VI).

STEP 7.

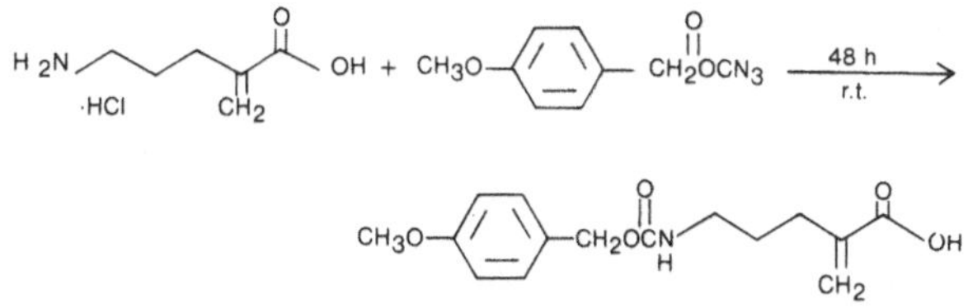

(VII)

To a solution of 3.8 g (53 mmol) of 2-methylene-5-aminopentanoic acid hydrochloride in 100 ml water was added 6.4 g (159 mmol) of magnesium oxide with stirring, followed by a solution of 12.2 g (59 mmol) of p-methoxybenzyloxycarbonyl azide in 100 ml of p-dioxane. The slurry was stirred at room temperature for two days. The mixture was filtered through hyflo and diluted with 200 ml ethylacetate. AG50W-X2 Dowex 50 ion-exchange resin (200 ml wet volume) was added and the mixture stirred at room temperature for 2 h. The resin was removed by filtration and washed with water. The filtrate layers were separated and the aqueous solution extracted twice with 200 ml of ethylacetate. The combined ethylacetate solution was dried over magnesium sulfate and concentrated in vacuo to give 15.5 g of 2-methylene-5-(p-methoxybenzyloxycarbonyl)-aminopentanoic acid (VII) as an oil.

STEP 8.

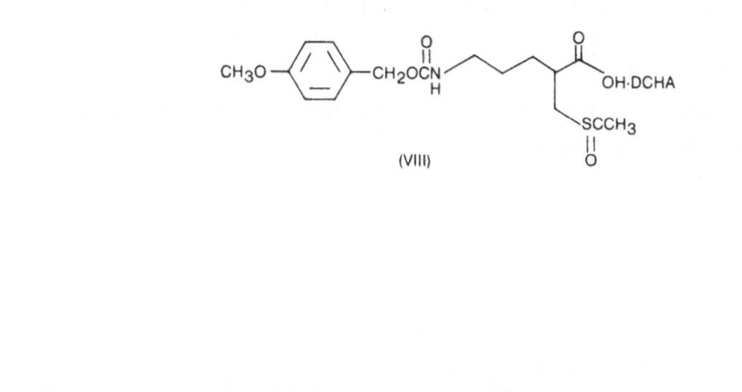

(VIII)

62

To 15.5 g of the oil containing the 2-methylene-5-(p-methoxybenzyloxycarbonyl)-aminopentanoic acid was added 50 ml of thiolacetic acid, and the solution allowed to stand at room temperature for 48 h. The thiolacetic acid was removed in vacuo (not heat) and three portions of benzene were added and removed in vacuo to remove excess thiolacetic acid. The resulting viscous oil was dissolved in ether (cloudy) and treated with a slight excess (10%) of dicyclohexylamine (DCHA). The temperature was lowered below 0 °C and crystals precipitated. Crystals were harvested to give 18.2 g of the dicyclohexyl-ammonium salt of 2-acetylthiomethyl-5-(p-methoxybenzyloxycarbonyl)-aminopentanoic acid (VIII).

STEP 9.

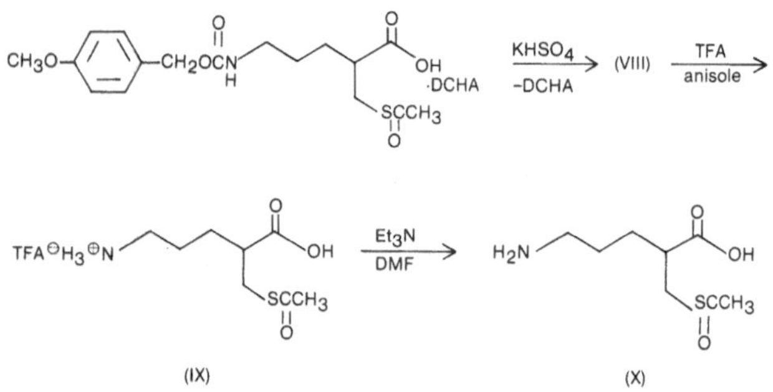

14.9 g (27 nmol) of the dicyclohexylammonium salt of VIII was dissolved in 200 ml chloroform and converted to the free acid by washing with two 200 ml portions of 10% potassium bisulfite solution. The chloroform solution was dried over magnesium sulfate and concentrated in vacuo. The residue was dissolved in 25 ml of anisole and cooled to 0 °C. To the cooled solution was added dropwise over 15 min, 125 ml of trifluoroacetic acid. After completion of addition, the solution was stirred at 0–5 °C for 1 h. The excess TFA was removed in vacuo. The residue was taken up in water and thoroughly washed with ether. The aqueous solution was lyophilized to a pale yellow oil. The oil was dissolved in 50 ml of dry DMF and treated with 10.3 ml triethylamine which resulted in precipitation of the 2-acetylthiomethyl-5-aminopentanoic acid (X). The residue was dried in vacuo, redissolved in water and lyophilized to 3.7 g of flocculent, off-white, solid.

STEP 10.

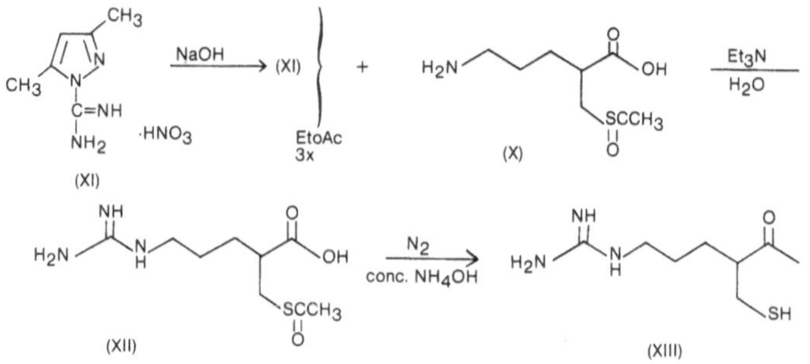

A 10 nmol (2.01 g) sample of 3,5-dimethyl-l-guanidopyrazole nitrate was dissolved in 100 ml 0.10 N NaOH and the free base extracted 3 × into 100 ml ethylacetate. The ethylacetate solutions were combined and dried over magnesium sulfate then concentrated in vacuo. The residue was dissolved in 2.6 ml water and 2 mmol (0.411 g) of 2-acetylthiomethyl-5-aminopentanoic acid was added, followed by 0.56 ml triethylamine. The solution was stirred at room temperature for three days. The mixture was diluted with water and washed with three two-fold volumes of ethylacetate. The aqueous solution was lyophilized. The residue was dissolved in 10 ml water and cooled to 0–5 °C under a nitrogen atmosphere for 3 h. The solution was diluted to 85 ml with water and the ammonia removed in acetate and lyophilized. The final product is (dl) 2-mercaptomethyl-5-guanidino-pentanoic acid (XIII) a potent inhibitor of serum carboxypeptidase N [EC 3.4.17.3] (11, 12).

Activation of human serum to generate C3a, C4a and C5a

We have previously described in detail procedures to activate human serum with inulin that generates C3a (13) or with zymozan (i.e., yeast cell ghosts) that generate both C3a and C5a (14). In these earlier studies the carboxypeptidase inhibitor 6-aminohexanoic acid (EACA)[1] was added to a final 1 M concentration in order to protect the anaphylatoxin from exopeptidase digestion during the course of complement activation.

---

[1] Abbreviations used in the text are EACA, epsilon-amino caproic acid or 6-amino hexanoic acid; SCPN, serum carboxypeptidase N.

Currently the approach in our laboratory has been to activate both complement pathways simultaneously thus generating all of the anaphylatoxins at one time. This is performed by first adding zymogen at 20 g/l of serum to generate C3a and C5a (as described by Fernandez and Hugli, 1976). After 10 min at 37 °C, heat-aggregated human IgG[2] was added to a final concentration of 0.5 g/l of serum and activation continued for an additional 30–35 min. Another variation from our earlier procedures involves selection of an inhibitor to prevent conversion of the anaphylatoxins in serum to the des Arg form. Instead of using the elevated concentrations of EACA, we presently use the potent serum carboxypeptidase inhibitor described by Ondetti, et al., (1979). The inhibitor 2-mercaptomethyl-5-guanidino-pentanoic acid was synthesized as previously described in the text and this compound, at a 1 mM concentration, is capable of protecting anaphylatoxins from conversion in serum for approximately one hour. These two changes from published activation protocols lead to a more facile isolation scheme for C3a and C5a, but also provides a source for the third anaphylatoxin C4a.

Isolation schemes for C3a, C4a and C5a

Step 1. The activated serum is acidified by adding 10 N HCl until the final concentration of HCl is 1 N. A major portion of the serum proteins precipitate during acidification and this fraction of insoluble protein is removed by centrifugation. The supernatant contains the three anaphylatoxins since they remain fully soluble and are functionally refractory to the acid treatment.

Step 2. The supernatant is dialyzed against 0.1 M ammonium formate at pH 4.0, lyophilized and dissolved in water to a final volume approximately one fifth that of the original supernatant.

Step 3. The dialyzed supernatant is gel filtered on a column of Bio Gel P-60 and fractions that contain anaphylatoxin are pooled and applied to a SP-Sephadex column. Alternatively, the dialyzed supernatant is absorbed by SP-Sephadex batch-

wise and eluted in one step by 0.8 M ammonium formate at pH 5.0. In this latter case, the effluent is dialyzed in 0.1 M ammonium formate at pH 5.0 before application to the column of SP-Sephadex (See Fig. 1). Anaphylatoxins elute from the SP-Sephadex in the order C5a, C4a and C3a under the influence of a linear salt gradient. At this point in the isolation scheme the anaphylatoxins are each purified separately using different chromatographic procedures. Whether Bio Gel P-60 or batchwise adsorption to SP-Sephadex is used depends on the volume of serum being processed. One liter or less of serum is easily filtered on the Bio Gel column; however a greater volume requires the batchwise technique.

Human C5a. The C5a fractions from SP-Sephadex are collected and the material is chromatographed over a column of CM-Sephadex eluted with 0.1 M ammonium formate at pH 7.0 according to the procedure of Fernandez and Hugli (14). If minor contaminants are observed after Microzone elec-

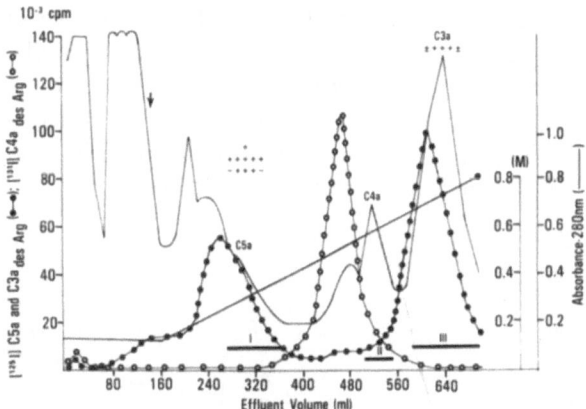

Fig. 1. Chromatography of the anaphylatoxin pool eluted from SP-Sephadex by 0.8 M ammonium formate at pH 5.0 after batchwise adsorption. The sample was dialyzed in 0.1 M NH₄CHO at pH 5.0 and applied to the SP-Sephadex column (1.6 × 30 cm) equilibrated in the pH 5.0 buffer. A majority of the contaminating serum proteins were eluted with a 0.1 M NH₄CHO buffer at pH 7.0 (150 ml). The anaphylatoxins were eluted at 60 ml/h from the column with a linear gradient from 0.1 to 0.8 M NH₄CHO at pH 7.0 (650 ml total volume). The effluent from the SP-Sephadex was monitored at 280 nm (——). ¹²⁵I labeled human C5a (●—●), ¹²⁵I labeled human C3a$_{des Arg}$ (●—●) and ¹³¹I labeled human C4a$_{des Arg}$ (○—○) were added and radioactivity of the effluent was also measured. Smooth muscle contracting activity identified the location of C3a and C5a. Pools were made of the effluent from the C5a, C4a and C3a regions of the chromatogram.

[2] Heat-aggregated IgG is prepared by exposing the protein in 0.02 M phosphate buffered saline at pH 7.2 to 64 °C for 20 min.

64

trophoresis on cellulose acetate strips at pH 8.6, a final step of chromatography is performed (15) on a QAE-Sephadex column.

Human C4a. The fractions from SP-Sephadex containing C4a are pooled and lyophilized. The C4a sample is dialyzed against 0.15 M ammonium formate at pH 5.5 (in small pore tubing (M. W. cut off 3,500)). This sample is applied to a CM-cellulose (CM-52) column and eluted with a gradient of 0.15 M to 0.35 M ammonium formate at pH 5.5 Fig. 2 illustrates the profile of C4a and ^{125}I labeled C4a$_{des\ Arg}$ eluted from this column. This separation is quite remarkable when one considers the difference between the two factors is only one formal positive charge.

Human C3a. A system analogous to that of C4a is used to further purify C3a obtained from the SP-Sephadex column. The C3a fractions are dialyzed and lyophilized as was C4a. The C3a is applied to a column of CM-Cellulose equilibrated in 0.15 ammonium formate at pH 7.0 and developed using a linear gradient from 0.15 M to 0.45 M in the salt (Fig. 3). Under these conditions of chromatography C3a and C3a$_{des\ Arg}$ are resolved and proper pooling of the sample can produce either factor in a homogeneous form as judged by electrophoresis at pH 8.6 on cellulose acetate strips (Fig. 4).

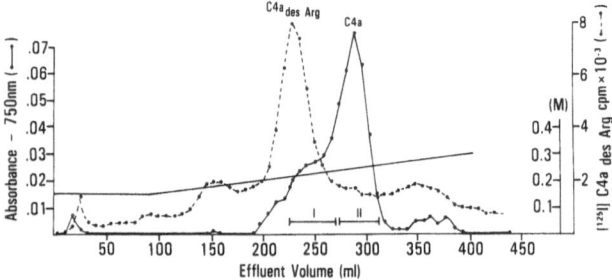

Fig. 2. Final purification step for human C4a. The C4a pool from SP-Sephadex was dialyzed in 0.15 M NH$_4$CHO at pH 5.5 and the sample was applied to a column (1.6 × 15 cm) of CM-cellulose (CM 52) equilibrated in the pH 5.5 buffer. A linear gradient of 0.15 M to 0.35 M NH$_4$CHO at pH 5.5 (total volume of 400 ml) was used to develop the column at a flow rate of 30 ml/h. The effluent was monitored by the Folin assay at 750 nm (●——●) and ^{125}I C4a$_{des\ Arg}$ was added to the sample as an internal reference (●——●). Human C4a elutes at a molarity (M) of 0.24 to 0.26 as a homogeneous protein and the des Arg derivative elutes just before the active parent molecule. Under these conditions the C4a or C4a$_{des\ Arg}$ are free of even trace contamination by C3a or C5a.

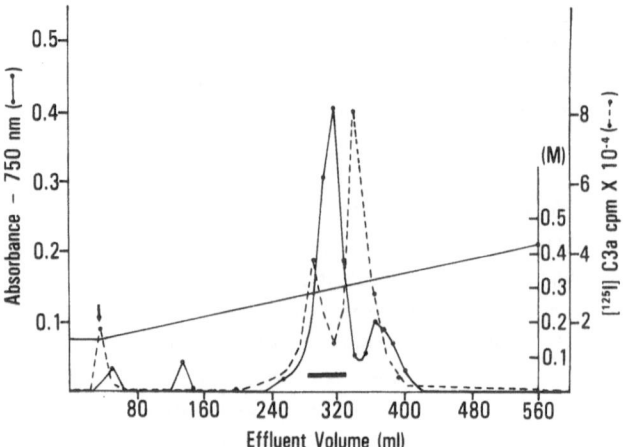

Fig. 3. Final purification step for human C3a. A pool of C3a$_{des\ Arg}$ from SP-Sephadex was dialyzed in 0.15 M NH$_4$CHO at pH 7.0 and the sample was applied to a column (1.5 × 27 cm) of CM-cellulose (CM 52) equilibrated in the pH 7.0 buffer. A linear gradient of 0.15 M to 0.45 M NH$_4$CHO at pH 7.0 (total volume of 600 ml) was used to develop the column at a flow rate of 30 ml/h. The effluent was monitored by the Folin assay at 750 nm (●——●). A sample oᶜ ^{125}I-human C3a was added to the C3a$_{des\ Arg}$ pool as an internal reference (●——●). The des Arg form of C3a elutes at a molarity (M) of 0.27 to 0.30 as a homogeneous protein. Under similar conditions C3a elutes at a molarity (M) of 0.31 to 0.34 (see shoulder after C3a$_{des\ Arg}$). C3a purified in this manner is devoid of either C4a or C5a.

Results and discussion

Effectiveness of carboxypeptidase inhibitors in protecting the anaphylatoxins from SCPN

A key to the generation of anaphylatoxins in serum is the effectiveness of the inhibitor that is selected to block the serum carboxypeptidase. This inhibitor must function without adversely affecting

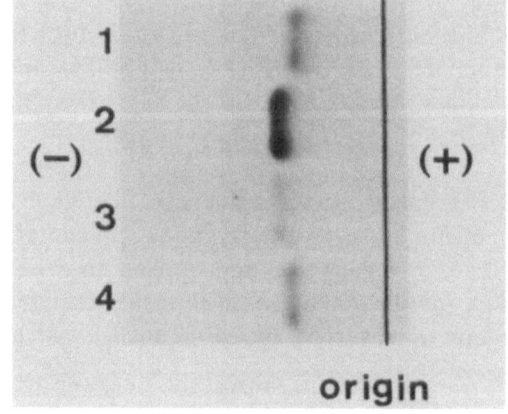

Fig. 4.

enzymes of either complement activation pathway. Previously, only the weakly inhibitory compound 6-aminohexanoic acid was available for use in preparative scale activations of serum. More recently there has been renewed interest in serum carboxypeptidase as an import control mechanism in anaphylatoxin action (16). Several high potency inhibitors of SCPN have now been designed beginning with the compound reported by Ondetti, et al., in 1979. Our synthesis of (dl) 2-mercaptomethyl-5-guanidinopentanoic acid is a variation of the procedure described previously (11) and the data in Table 1 indicates the K_i of this inhibitor for SCPN is only slightly greater than that of 2-mercaptomethyl-3-guanidinoethylthiopropanoic acid, an inhibitor described recently by Plummer and Ryan (12). There is every indication that the inhibitor described by Plummer and Ryan may be easier to synthesize and can be obtained in higher yields. It is now obvious from K_i measurements with purified SCPN that either of the mercapto-guanidino compounds are far superior to the dicarboxylic-guanidino inhibitors of McKay, et al. (17), and are 10^6 times more effective than EACA for inhibiting purified SCPN. Our interest, of course, is whether these inhibitors can function effectively in serum. Measurements of anaphylatoxin activity in serum in the presence of the various inhibitors indicate that either of the mercaptoguanidino compounds at 1 mM will prevent inactivation of human C3a in serum for up to one hour. It was known from previous reports that 1 M 6-aminohexanoic acid is required to offer the same level of protection.

When we used 1 mM 2-mercaptomethyl-5-guanidinopentanoic acid (dl) in human serum less than 20% of the C3a was converted to C3a$_{des Arg}$ after 45 min of complement activation at 37 °C. Since C3a is the most susceptible of the anaphyla-

toxin molecules to SCPN, we anticipate that even better protection exists for C4a and C5a.

Chromatographic separation of human C3a, C4a and C5a

The procedures described in the text provide the rigorous purification techniques that have established the following facts: 1) that the des Arg form of human C5a retains weak intrinsic biological activity (18); 2) that human C3a devoid of C5a exhibits no chemotactic activity for human neutrophils in the physiologic concentration range (6); and more recently 3) that C4a activity is not an artifact of contaminating levels of C3a. Another major benefit of our refined methodology for anaphylatoxin isolation is that it permits one to obtain each factor absolutely free from trace levels of any other anaphylatoxin. This becomes particularly important when the anaphylatoxins are being used as antigens in raising monospecific antibodies or as standards in sensitive radioimmune assay procedures (19). Finally, the refined isolation procedures should find utility in preparing the individual anaphylatoxins in their fully active forms since gradient elution, in each specific case, resolves the intact factor from the des Arg derivative.

If one analyzes the data in Fig. 1 carefully it becomes obvious that during an initial isolation of these factors a bioassay is required to detect the individual anaphylatoxins. A reliable assay for this purpose is the guinea pig ileal assay that detects these factors by their ability to contract the tissue (20). Each factor has a unique specific activity in this assay procedure and the order of relative activities are C5a > C3a > C4a (Table 2). Once the factors have been purified to homogeneity, it is possible to radioiodinate them and the labelled

Table 1. Inhibitors of serum carboxypeptidase N.[a]

Inhibitor	K_i (M)
2-Mercaptomethyl-5-guanidinopentanoic acid (d, l)	1.0×10^{-9}
2-Mercaptomethyl-3-guanidinoethylthiopropanoic acid (dl)[b]	2.0×10^{-9}
Guanidinoethylthiosuccinic acid (dl)[b]	1.0×10^{6}
6-Aminohexanoic acid (ε-aminocaproic acid)	2.0×10^{-3}

[a] Serum carboxypeptidase N was purified according to the procedure of Plummer and Hurwitz, 1978. J. Biol. Chem. 253: 3907–3916.

[b] Taken from Plummer and Ryan (12).

Table 2. Specific activity of the human anaphylatoxin on guinea pig ileum.

	Dosage for smooth muscle contraction	Relative activity	Tachyphylactic for C3a	C4a	C5a
C3a	6 to 9 × 10^{-9} M	200	+	+	-
C4a	1 to 2 × 10^{-6} M	1	+	+	-
C5a	4 to 6 × 10^{-10} M	3 000	-	-	+

Terminal strips of guinea pig ileum contract when exposed to the above concentrations of each anaphylatoxin. The mechanism in this particular tissue is believed to be almost exclusively histamine mediated.

factors are used to monitor all future preparations. This approach avoids the necessity of carrying out time-consuming bioassays until the factor is purified. The bioassay is capable of distinguishing between C5a and either C3a or C4a because the tissue has independent receptors for C5a and C3a. On the other hand, C3a and C4a cannot be distinguished in qualitative terms using the ileal assay because they cross-desensitize the tissue to one another (2).

Perhaps the broadest implications of having isolation procedure that provide an highly purified anaphylatoxin will be in the area of diagnostic medicine. Preliminary evidence already supports this contention since patterns of anaphylatoxin generation in the blood of patients experiencing problems as diverse as hypersensitivity (21) and corrective cardiosurgery (22) may actually reflect the severity of the clinical problem. There must be many other pathological conditions involving complement activation that could be monitored by detecting and quantitating anaphylatoxin generation. Such studies are possible only when suitably pure reagents are available to insure valid and accurate results.

Acknowledgments

This research was supported in part by PHS grants HL25658 and HL16411 from the Heart, Lung and Blood Institute of the National Institutes of Health. Support was also provided by Eli Lilly and Co. The authors wish to thank Miss Holly Wimer for her assistance in preparing the manuscript. This is publication number 2483 from the Research Institute of Scripps Clinic.

References

1. Hugli, T. E., 1981. CRC Critical Reviews in Immunology. 1: 321–366.
2. Gorski, J. P., Hugli, T. E. & Müller-Eberhard, H. J., 1979. Proc. Natl. Acad. Sci. USA. 76: 5299–5302.
3. Gorski, J. P., Hugli, T. E. & Mülle-Eberhard, H. J., 1981. J. Biol. Chem. 256: 2707–2711.
4. Hugli, T. E. & Müller-Eberhard, H. J., 1978. Adv. Immunol. 26: 1–53.
5. Jensen, J. A., 1972. In Biological Activities of Complement, (Ingram, D. G. Ed.), S. Krager, Basel, pp. 136–157.
6. Fernandez, H. N., Henson, P. M., Otani, A. & Hugli, T. E., 1978. J. Immunol. 120: 109–115.
7. Stimler, N. P., Brocklehurst, W. E., Bloor, C. M. & Hugli, T. E., 1980. J. Pharm. Pharmacol. 32: 804.
8. Stimler, N. P., Brocklehurst, W. E., Bloor, C. M. & Hugli, T. E., 1981. J. Immunol., in press.
9. Orr, W., Varani, J. & Ward, P. A., 1978. Am. J. Pathol., 93: 405–422.
10. Orr, W., Phan, S. H., Varani, J., Ward, P. A., Kreutzer, D. L., Webster, R. O. & Henson, P. M., 1979. Proc. Natl. Acad. Sci. USA, 76: 1986–1989.
11. Ondetti, M. A., Condon, M. E., Reid, J., Sabo, E. G., Cheung, H. S. & Cushman, D. W., 1979. Biochemistry 18: 1427–1430.
12. Plummer, T. H. Jr. & Ryan, T. J., 1981. Biochem. Biophys. Res. Commun. 98: 448–454.
13. Hugli, T. E., Vallota, E. H. & Müller-Eberhard, H. J., 1975. J. Biol. Chem. 250: 1472–1478.
14. Fernandez, H. N. & Hugli, T. E., 1976. J. Immunol. 117: 1688–1694.
15. Gerard, C. & Hugli, T. E., 1979. J. Biol. Chem. 254: 6346–6351.
16. Mathews, K. P., Pan, P. M., Gardner, H. J. & Hugli, T. E., 1980. Annals Intern. Med. 93: 443–445.
17. McKay, T. J. & Plummer, T. H. Jr., 1978. Biochemistry 17: 401–405.
18. Gerard, C. & Hugli, T. E., 1981. Proc. Natl. Acad. Sci., USA 78: 1833–1837.
19. Hugli, T. E. & Chenoweth, D. E., 1980. In Immunoassays: Clinical Laboratory Techniques for the 1980's, Alan R. Liss, Inc., New York, NY pp. 443–460.
20. Cochrane, C. G. & Müller-Eberhard, H. J., 1968. J. Exp. Med. 127: 371–386.
21. Curd, J. G., Chenoweth, D. E. & Hugli, T. E., 1980. J. Immunol. 124: 1517 (Abstract).
22. Chenoweth, D. E., Cooper, S. W., Hugli, T. E., Stewart, R. W., Blackstone, E. H. & Kirklin, J. W., 1981. N. Engl. J. Med. 304: 497–503.

Received May 29, 1981.

The serum thymic factor (FTS)
Chemical and biological aspects

Jean-Marie Pleau, Mireille Dardenne and Jean-François Bach
Inserm U 25, Hôpital Necker, 75015 Paris, France

Summary

The thymus produces a nonapeptide capable of inducing T cell surface markers and T cell functions in immature lymphoid cells. This peptide is found partly bound to a carrier protein in the circulation from where it has been isolated (hence its name of serum thymic factor (FTS)). Immunofluorescence studies using an antibody raised against synthetic FTS has shown that it is produced by the thymic epithelium. Its mode of action at the cellular level involves the binding to specific high affinity receptors.

Introduction

The thymus and its cellular products, the thymus-derived cells (T cells) play a central role in the generation of effector cells in cell-mediated immunity and in the regulation of the various categories of immune responses. The thymus produces polypeptides, which induce lymphocyte differentiation *in vitro* and *in vivo*. Several thymic peptides have been isolated and eventually characterized and it appears that there is probably more than one biologically active thymic factor. Some peptides have been isolated from thymic tissue, others have also been obtained from serum. Before reviewing the serum thymic factor which we have characterized, we shall summarize the data available on the other main preparations issued from calf thymus: thymosin, thymopoietin and thymic humoral factor.

Thymus extracted preparations

A. Goldstein extracted from calf thymus gland a crude fraction called *thymosin* Fraction V (1). This extract, which contains a whole set of molecules,

has been shown to be biologically active in various assay systems including an MLR (mixed lymphocyte reaction) assay measuring the proliferation of murine thymocytes in an allogeneic mixed leukocyte culture (2), a human E-rosette assay (3), and mouse mitogen assays *in vivo* (4). Moreover, thymosin Fraction V has been found to induce T cell differentiation and enhance immunological functions in genetically athymic nude mice, in adult thymectomized (ATx) mice, in NZB mice with severe autoimmune reactions (5) and in thymus-deprived and/or immunodeprived individuals (6). More recently, thymosin Fraction V has also been shown to induce terminal deoxynucleotidyl transferase (TdT) in the less dense layers of bone marrow cells of nude and of ATx mice (7).

Thymosin Fraction V is a partially purified preparation. Some of its polypeptidic components have been characterized. These peptides have been divided into three categories, based on their isoelectric pattern. The first polypeptide isolated from the highly acidic region of Fraction V has been termed 'thymosin α1'. Its primary structure has recently been completely elucidated (8). This peptide is active in several bioassay systems, but apparently only possesses a minor part of the total

Molecular and Cellular Biochemistry 41, 67–72 (1981).   0300-8177/81/0041-0067/$01.20.

biological activity of Fraction V: thymosin $\alpha 1$ was recently found to potentiate helper cells in vitro, while another peptide isolated from thymosin, $\alpha 7$ was found to induce suppressor cells (9).

Thymopoietin, another polypeptide factor of the thymus isolated by G. Goldstein, induces the differentiation of prothymocytes into thymocytes, and also has secondary effects on neuromuscular transmission (10). Thymopoietin is a polypeptidic chain made of 49 amino acids, whose sequence has been completely determined. Two forms of bovine thymopoietin (I and II) were isolated; they differ in their amino acid sequence by 3 amino acids (11). By peptide synthesis, a tridecapeptide based on residues 29–41 (12), and a pentapeptide (TP5) corresponding to the residues 38–36 of thymopoietin (13) have been shown to have effects similar to those of thymopoietin itself. Thymopoietin or its shorter analogues have been shown to induce T cell markers and to promote several T cell functions such as T cell mediated cytotoxicity and suppressor T cells (14).

Using an in vitro graft versus host (GVH) assay, Trainin *et al.,* (15) have characterized a small molecular weight peptide called *'thymic humoral factor'* (THF). THF was isolated by a procedure which consisted of a stepwise gel filtration through Sephadex columns and assayed in terms of the acquisition of competence by spleen cells from neonatally Tx mice in an in vitro assay of GVH reactivity. It appeared that THF is a polypeptide of 3000 mol wt which migrates as a single band in isoelectrofocusing. Amino acid analysis of acid hydrolysates revealed that purified THF contains approximately 30 residues (15). THF is capable of restoring the competence of lymphoid cells from neonatally Tx mice to participate in MLR (16), to kill tumor cells (17), and to react to T lectins (18). Immune maturation of thymus-derived lymphoid cells after *in vitro* or *in vivo* exposure to THF occurs via an obligatory rise in cellular cAMP level (19).

Serum thymic factor (FTS for Facteur Thymique Sérique)

Our approach to the problem was different. We had initially shown that thymic extracts (prepared by A. L. Goldstein & N. Trainin) conferred sensitivity to inhibition by anti-theta serum (AθS) and azathioprine (Az) on spleen rosette forming cells (RFC) from ATx mice, which normally lack such properties (20). These data were later confirmed by Komuro & Boyse, employing direct cytotoxicity assays on spleen cells fractionated on albumin gradients (21). Subsequently, we reported that normal mouse and pig serum contained an active principle which had the same activity on θ-negative spleen RFC as thymic extracts (22). This serum thymic factor, called Facteur Thymique Sérique (FTS) has been isolated from pig serum. It is a nonapeptide (23), the smallest of the thymic peptides isolated so far in an apparently natural state. The thymus dependency of FTS was ascertained by its absence in the serum of Tx and nude mice and its reappearance after grafting a thymus gland or pure thymic epithelial cells in the serum (24). Moreover, we have recently presented evidence of the presence of FTS in the thymus, including demonstration, in the rosette assay, of a material of identical biological activity and similar mol wt as FTS, and cross-reactivity with an antibody raised in rabbits against FTS (25). Lastly, the recent observation of the localization of these antibodies on the thymic epithelium has provided direct demonstration of their thymic origin (26).

FTS has been sequenced and synthesized (27) and synthetic FTS has the same biological and chemical characteristics as natural FTS. Recently, we demonstrated the existence of a FTS-specific receptor on continuous human T cell lines (28).

Isolation and purification (29)

Preparative studies were done using pig serum. Defibrinated pig serum (1000 l) was ultrafiltered on an hemodialyzer RP6 Rhône Poulenc. Four hundred and twenty liters of serum ultrafiltrate were concentrated 200 times on Amicon membranes up to 2.1 l. Amicon concentrates were submitted to the four following steps: Sephadex G-25 filtration in neutral medium, carboxymethyl cellulose (CMC) chromatography, Sephadex G-25 in acetic acid medium, and Sephadex G-10 in distilled water. Active fractions were detected by the rosette assay.

The method used has already been described (30). In brief, fractions were tested for their ability to confer on spleen RFC from ATx C57BL/6 mice the sensitivity to Az that they were lacking. The fractions were incubated for 90 min at 37 °C with spleen cells from C57BL/6 mice thymectomized 10–15 days before. Az (10 μg/ml) was added simultaneously with the serum. At the end of the incubation, 12×10^6 sheep red blood cells (SRBC) were added to the cells which were then centrifuged and resuspended by rotation on a roller (10 cm diameter) at low speed (10 rpm). RFC were counted in a hemocytometer at low magnification ($\times 250$).

The fraction containing FTS activity, obtained after such filtration, gave a reproducible amino acid analysis (lysine, aspartic acid, serine 2, glutamic acid 2, glycine 2, alanine). No free amino acids were found in the unhydrolized samples thus obtained and dansylation performed on different batches was negative, no dansyl derivative could be identified. In view of these results it appeared highly probable that FTS had a blocked NH_2-terminal end. To confirm this hypothesis a dansyl-negative sample was treated with pyrrolidone carboxy peptidase: one single dansyl derivative, i.e. alanine was identified.

Sequence studies (27)

As a result of amino acid analysis and sequence studies on the intact peptide and on the peptide treated with proteolytic enzymes by Edman's method, the amino acid sequence proposed for FTS was the following:

<Glu–Ala–Lys–Ser–Gln–Gly–Gly–Ser–Asn–OH

There is no apparent species specificity since the amino acid analysis of calf and human FTS is identical to that of porcine FTS. This sequence does not show any homology with that of the other thymic peptides that have also been sequenced (thymopoietin, thymosin α1). One cannot exclude, however, that peptides not chemically related to FTS serve as cleavage factor for an FTS precursor, as is known to be the case for growth hormone which induces the release of small peptides, the somatomedins that mediate most of its biological activities.

Synthesis

On the basis of this sequence, a peptide has been synthesized according to two methods: using solid phase synthesis (Merrifield's technique) by the Merck peptide group (Drs. Veber & Hirschman) and classical solution methods (Dr. Bricas and Choay's peptide group, Dr. Lefrancier) (31). The synthetic material showed full biological activity and displayed chromatographically characteristics identical to those of biological FTS in several chromatography systems (27). Recently, a series of analogue molecules was synthesized by Dr. Bricas (32) which made it possible to identify the biologically active site and the antigenic site of the molecule (33).

Inhibitors and carriers of FTS

Several observations argue in favor of the existence in the serum of FTS inhibitors detectable in the rosette assay. Total mouse serum examined in the test described above does not possess the activity reported for serum ultrafiltrate. This activity only appears after elimination of molecules with mol wt between 100 000 and 300 000 as assessed by dialfiltration on Amicon membrane (33). Therefore, there seems to exist an FTS inhibitor with high mol wt present in normal serum and separable from FTS by simple dialysis or dialfiltration. The molecule responsible for this inhibitory activity of normal serum is not known, but it has been shown to be active at very low concentrations since total serum diluted 1/5000 keeps its inhibitory activity (33).

Moreover, when ultrafiltered serum is concentrated and chromatographed on G-25 Sephadex, the activity of the active fractions is increased relative to the initial sample, which is very suggestive of low mol wt (ultrafiltrable) inhibitors. Using an inhibitor assay identical to that described above, we have further characterized these inhibitors. Pig serum has been fractionated on G-25 Sephadex (phosphate buffer) and the inhibitory fractions found in the void volume have been further fractionated on G-50 and G-100 Sephadex columns. Two inhibitors could thus be characterized with approximative mol wt of 4000 and 20 000 respectively (33).

We have also described the presence in the serum of FTS carrier molecules (34). Such demonstration was obtained by fractionation of human or mouse serum on G-150 and evaluation of the eluted fractions in the rosette assay, which revealed the presence of two peaks of activity. One peak corresponded to the elution volume of FTS (mol wt 867), the other one to molecules with mol wt of 40 000–60 000 daltons. Both peaks were absent in Tx mouse serum but could be induced in such Tx mice by injection of synthetic FTS. A study of the biological half-life of both peaks (assessed by the rosette assay) and a demonstration of their specific retention on anti-FTS immunoadsorbent along with direct binding experiments using tritium-labelled FTS altogether suggest that FTS is transported in the serum by a molecule with mol wt of the order of that of albumin or prealbumin, and that FTS bound to this molecule is responsible for the biological FTS-like activity associated with large mol wt fractions of normal serum.

The relationship between such still ill-defined carrier proteins and prealbumin, reported by Burton et al. (35) to be active in the rosette assay, is a matter of speculation but should be taken into consideration since 1) the prealbumin level is not known to be decreased in Tx mice or aged humans where it shows no biological activity, 2) prealbumin is a known carrier of a number of small molecules and 3) passage of 'albumin zone' fraction of Sephadex G-150 separated serum fractions on anti-FTS immunoadsorbent free of any anti-BSA antibody activity) removes the totality of serum rosette activity.

Biological activities

FTS has been isolated on the basis of the rosette assay described above. Used either as the natural or synthetic form, it has also been proven to be active, as other purified or synthetic peptides, on most T cell markers and functions.

Marker studies indicate that various types of T cell differentiation antigens may be induced in precursor cells that are devoid of such markers. This was initially shown by us for the theta antigen, and has now been confirmed and extended for Ly antigens in the mouse and xenogeneic T cell antigens in the human (36). Moreover, synthetic FTS has been shown to normalize the abnormally high levels of autologous erythrocyte-binding cells in ATx mice (37), and increased E-rosette numbers in patients with T cell deficiencies. TdT expression is decreased in normal bone marrow cells by synthetic FTS and in BSA gradient-separated human bone marrow cells (36).

FTS has been shown to be active on most T cell functions. It enhances T cell mediated cytotoxicity in Tx mice. This effect is particularly clear in ATx mice using the Brunner assay (38). It is not known whether FTS directly stimulates the generation of the cytotoxic cells or enhances the function of a regulatory cell that could be the adult thymectomy-sensitive Ly 123^+ spleen cell. Similarly, FTS also acts on the T cells involved in delayed-type hypersensitivity induced by DNFB (39). It restores a normal response in ATx mice. Its effect on helper T cells as studied on anti-SRBC antibody production is much less clear, perhaps due to a simultaneous action on suppressor T cells. In fact, FTS has recently proven to be remarkably active on suppressor T cells in various in vitro and in vivo systems. Given in vivo to normal mice, FTS suppresses the generation of alloantigen reactive T cells or DNFB-sensitive T cells (39). Given at 10–100 ng it may prolong skin allograft survival (33) or enhance the growth of MSV-sarcoma in T cell deprived mice (40) (while at lower doses it stimulates its rejection).

It is interestig to note that this effect of FTS on suppressor cells probably explains most of its preventive effects observed in NZB autoimmune mice (such as the decrease in anti-PVP antibody production or the prevention of SJÖGREN syndrome (40). A simultaneous effect on helper T cells probably explains the accelerated production of IgG anti-DNA antibodies also observed in these mice (40).

Other biological effects are less well explained, such as the stimulation of CFU-C (41) or the enhancement of NK cell activity recently observed in mice and in man (42).

Finally, FTS shows a large variety of effects on T cell functions, especially in relatively mature cells, including suppressor T cells. In addition to its effect of T cell maturation, it seems to have a pharmacological effect on mature T cells, and particularly suppressor cells that would still possess receptors for it.

71

Radioimmunoassay (RIA)

Using an antiserum raised in rabbits against BSA-coupled FTS, a RIA for FTS has been set up (33). An *anti-FTS antibody* was obtained by coupling FTS to BSA in the presence of glutaraldehyde which binds to the amino groups of FTS and BSA. Rabbits were immunized with BSA-FTS conjugate intradermally in several places, in the presence of complete Freund's adjuvant (1 mg total FTS-BSA conjugate). Three months later, the rabbits were boosted with 0.5 mg BSA-FTS conjugate. Bleeding was performed each week and serum samples assays by a Farr test using ^{125}I-labelled FTS. Peak antibody titers were obtained one month after boosting (1/300 000).

Anti-FTS antibodies were characterized by their capacity to bind ^{125}I-labelled FTS and related structural analogues. Details of this assay were described elsewhere (43). FTS was ^{125}I-labelled by use of the Bolton Hunter reagent. Binding studies were performed by incubating labelled FTS (5 000 cpm) and graded dilutions of serum samples in phosphate buffer 0.1 M pH 7.3 overnight at 4 °C. Separation of free and antibody-bound FTS was obtained by propanol precipitation.

The anti-FTS serum-bound labelled FTS and the binding were totally displayed by addition of cold FTS (2×10^{-10} M) but not by a close structural analogue with Asp instead of Asn, as the terminal residue.

The application of RIA to serum samples has been confronted with several complications: low levels of serum FTS, extreme sensitivity of FTS to proteolysis, and existence of serum-interfering proteins. The demonstration of immunoreactive FTS in normal serum and its absence in the serum of Tx animals has been done using Amicon filtration, concentration and G-25 Sephadex chromatography, but this procedure was impossible to use routinely.

Specific cellular receptors for FTS

The first step in the action of a polypeptide hormone is the binding to specific receptors on its target cell. It is thus of utmost importance toward the understanding of the cellular mode of action of FTS to search for and characterize FTS receptors on lymphoid cells. To do this, we have applied the methodology that has been established for polypeptide hormones.

We have demonstrated the presence of specific FTS receptors on a T cell line derived from a patient with acute lymphoblastic leukemia (1301) (28). FTS was labelled with tritium, using sodium borohydride. Cultured lymphocytes were incubated with labelled FTS for 90 min; 12 to 15% FTS bound to the cell line and the specificity of the binding was assessed by the strong inhibition of the binding obtained by addition of unlabelled FTS (10^{-5} M). In fact, it was even possible to displace labelled FTS by secondary addition of unlabelled FTS after 90 min incubation. The production of FTS receptors by the T cell line was verified by their disappearance after trypsin treatment and their reappearance after overnight incubation.

The capacity of the 1301 cell line to bind various peptides, related or not to FTS, was investigated using the same methodology as described above for FTS. Ocytocin and angiotension (10^{-8} to 10^{-5} M), two peptides unrelated to FTS but of similar size, did not show any significant displacement of ^3H FTS binding. The 1301 cell line binds human insulin, as has been previously reported for the cell line IM9 (42). Such insulin receptors were shown to be independent from FTS receptors, since addition of unlabelled insulin induced the inhibition of the binding of ^{125}I-labelled insulin, but not ^3H FTS binding, and conversely, the presence of unlabelled FTS (10^{-5} M) left ^{125}I-insulin binding unaltered, whereas strong inhibition of ^3H FTS occurred.

The specificity of the cell receptors for FTS was further confirmed by experiments studying the inhibition of ^3H FTS binding by low concentrations of FTS analogues (heptapeptide and octapeptide or FTS analogues including a D amino acid). Minor changes in the FTS sequence, that do not alter biological activity alter ^3H FTS binding to 1301 but, conversely, analogues that are not fully active in the rosette assay are less effective competitors compared to original FTS. It is interesting to note that another T cell was shown to bind FTS, whereas four B cell and one null cell lines were negative. The fact that some T cell lines did not bind FTS indicates that T cells do not express FTS receptors at all stages of T cell differentiation. Preliminary data have shown that normal human peripheral blood lymphocytes express FTS receptors when they are preincubated in medium, probably to let them shed already bound FTS. Finally, receptors with a high binding affinity for FTS ($K_D = 10^{-9}$ M) appear to exist on T cells. That these receptors are associated with the biological effects of FTS is likely but remains to be demonstrated.

References

1. Hooper, J. A., Mc Daniel, M. C., Thurman, G. B., Cohen, G. H., Schulof, R. S. & Goldstein, A. L., 1975. Ann. N.Y. Acad. Sci. 249: 125–138.
2. Cohen, G., Hooper, J. A. & Goldstein, A. L., 1975. Ann. N.Y. Acad. Sci. 249: 145–153.
3. Wara, D. W., Goldstein, A. L., Doyle, W. & Amman, A. J., 1975. N. Engl. J. Med. 292: 70–74.
4. Thurman, G. B. & Goldstein, A. L., 1975. In: 'The biological activity of thymic hormones' Van Bekkum, D. W., ed.), pp. 261–264, Kooyker Sci. Publ., Rotterdam.
5. Thurman, G. B., Ahmed, A., Strong, D. M., Gershwin, M. E., Steinberg, A. D. & Goldstein, A. L., 1975. Transpl. Proc. 7: 299–303.
6. Horowitz, S. W., Borcherding, A. V., Morrthy, R., Chesney, H., Schulte-Wisserman, R., Hong, C. & Goldstein, A. L., 1977. Science 197: 999–1001.
7. Pazmino, N. H., Ihle, J. N. & Goldstein, A. L., 1978. J. Exp. Med. 147: 708–718.
8. Goldstein, A. L., Low, T. L. K., Mac Adoo, M., Mc Clure, J., Thurman, G., Rossio, J. Lai, C., Chang, D., Wang, S., Harvey, C., Ramel, A. H. & Meinhofer, J., 1977. Proc. Nat. Acad. Sci. (Wash) 74: 725–729.
9. Low, T. L. K., Thurman, G. B., Chincarini, C., Mc Clure, J. E., Marshall, G. D., Hu, S. K. & Goldstein, A. L., 1979. Ann. N.Y. Acad. Sci. 332: 348.
10. Goldstein, G., 1975. Ann. N.Y. Acad. Sci. 249: 177–185.
11. Schlesinger, D. H. & Goldstein, G., 1975. Cell 5: 361–365.
12. Schlesinger, D. H., Goldstein, G., Scheid, M. & Boyse, E. A., 1975. Cell 5: 367–371.
13. Goldstein, G., Scheid, M., Boyse, E. A., Schlesinger, D. H. & Van Wauwe, J., 1979. Science 204: 1309–1310.
14. Law, C. & Goldstein, G., 1980. J. Immunol. 124: 1861–1865.
15. Trainin, N., Small, M., Zipori, D., Umiel, T., Kook, A. I. & Rotter, V., 1975. The biological activity of thymic hormones (Van Bekkum, D. W., ed.), pp. 117–144, Kooyker Sci. Publ., Rotterdam.
16. Umiel, T. & Trainin, N., 1975. Eur. J. Immunol. 5: 85–88.
17. Carnaud, C., Ilfeld, D., Brook, I. & Trainin, N., 1973. J. Exp. Med. 138: 1521–1532.
18. Rotter, V. & Trainin, N., 1975. Cell. Immunol. 16: 413–421.
19. Kook, A. I. & Trainin, N., 1974. J. exp. Med. 139: 193–207.
20. Bach, J. F., Dardenne, M., Goldstein, A., Guha, A. & White, A., 1971. Proc. Natl. Acad. Sci. U.S.A. 68: 2734–2738.
21. Komuro, K. & Boyse, E. A., 1973. Lancet 1: 740–743.
22. Bach, J. F., & Dardenne, M., 1973. Immunology 25: 353–366.
23. Bach, J. F., Dardenne, M. & Pléau, J. M., 1977. Nature 266: 55–56.
24. Dardenne, M., Papiernik, M., Bach, J. F. & Stutman, O., 1974. Immunology 27: 299–304.
25. Dardenne, M., Pléau, J. M., Blouquit, Y. & Bach, J. F., 1980. Clin. Exp. Immunol. 42: 477–482.
26. Monier, J. C., Dardenne, M., Pléau, J. M., Schmitt, D., Deschaux, P. & Bach, J. F., 1981. Clin. Exp. Immunol 42: 470–476.
27. Pléau, J. M., Dardenne, M., Blouquit, Y. & Bach, J. F., 1977. J. Biol. Chem. 252: 8045–8047.
28. Pléau, J. M., Fuentes, V., Morgat, J. L. & Bach, J. F., 1980. Proc. Natl. Acad. Sci. U.S.A. 77: 2861–2865.
29. Dardenne, M., Pléau, J. M., Man, N. K. & Bach, J. F., 1977. J. Biol. Chem. 252: 8040–8044.
30. Dardenne, M. & Bach, J. F., 1975. The Biological activity of thymic hormones (Van Bekkum, D. W., ed.), pp. 235–243, Kooyker Sci. Publ., Rotterdam.
31. Bricas, E., Martinez, J., Blanot, D., Auger, G., Dardenne, M., Pléau, J. M. & Bach, J. F., 1977. 'Peptides', Proc. 5th Amer. Peptide Symposium (Goodman, M. & Meinhofer, J., eds.) pp. 564–567, J. Wiley & Sons.
32. Blanot, D., Martinez, J., Auger, G. & Bricas, E., 1979. Int. J. Peptide Protein Res. 14: 41–56.
33. Bach, J. F., Bach, M. A., Blanot, D., Bricas, E., Charreire, J., Dardenne, M., Fournier, C. & Pléau, J. M., 1978. Bull. Inst. Pasteur 76: 325–398.
34. Dardenne, M., Pléau, J. M. & Bach, J. F., 1980. Eur. J. Immunol. 10: 83–86.
35. Burton, P., Iden, D., Mitchel, K. & White, A., 1978. Proc. Natl. Acad. Sci. U.S.A. 75: 823–827.
36. Incefy, G., Mertelsmann, R., Iwata, K., Dardenne, M., Bach, J. F. & Good, R. A., 1980. Clin. Exp. Immunol. 40: 396–406.
37. Charreire, J. & Bach, J. F., 1975. Proc. Natl. Acad. Sci. U.S.A. 72: 3201–3205.
38. Bach, M. A., 1977. J. Immunol. 119: 641–647.
39. Erard, D., Charreire, J., Auffredou, M. T., Galanaud, P. & Bach, J. F., 1979. J. Immuol. 123: 1573–1576.
40. Bach, M. A., Fournier, C. & Bach, J. F., 1979. In: '12th International Leukocyte Conference' (Quastel, M. R., ed.), pp. 177–188, Academic Press, New York.
41. Lepault, F., Dardenne, M. & Frindel, E., 1979. Eur. J. Immunol. 9: 661–664.
42. Bardos, P., Carnaud, C. & Bach, J. F., 1980. C.R. Acad. Sci. Paris 289: Série D., 1251–1254.
43. Pléau, J. M., Pasques, D., Bach, J. F., Gros, C. & Dray, F., 1977. Radioimmunoassay and related procedures in medicine, Vol. II, pp. 505–510, I.A.E.A. Vienna.
44. Lesniak, M., Gordon, P., Roth, J. & Gavin, J., 1974. J. Biol. Chem. 249: 1664–1668.

Received November 3, 1980.

Tuftsin, Thr-Lys-Pro-Arg
Anatomy of an immunologically active peptide

Mati Fridkin and Philip Gottlieb
Dept. of Organic Chemistry, The Weizmann Institute of Science, Rehovot, Israel

Summary

Tuftsin, a natural occurring tetrapeptide, has been found to exhibit several biological activities connected with immune system function. Although little is known about tuftsin's 'biogenesis', much information has been gleaned about its structure-function relationships, which have shown that several features of the molecule are essential for expression of full biological activity. Furthermore, specific receptor sites for tuftsin have been found to exist exclusively on phagocytic cells. Research indicates that tuftsin binding to target cells effect intracellular calcium and cyclic nucleotide levels. Implication of these facts on tuftsin's mode of action are discussed.

Basic peptidic segments resembling tuftsin are found in a variety of regulatory peptides. Questions are, therefore, raised as to the biospecificity and cross-reactivity of these sequences. Substance P, one such peptide, which binds with and activates tuftsin receptors, is described.

In light of tuftsin's therapeutic potential, assays for its determination have been introduced. When applied to analyze human blood serum of normal as well as of various pathological origins, direct correlation was found between tuftsin levels and susceptibility to bacterial infections.

Introduction

The discovery of the natural occurring peptide tuftsin was the culmination of studies by Najjar and coworkers on the physiological role of cytophilic γ-globulins (1-6). In that series of communications, it was demonstrated that particular fractions of the immunoglobulins, from mammalian origin, obtained through chromatography on phosphocellulose columns, have selective affinities to different blood cells: the erythrocyte (2, 6), the thrombocyte (6, 7) and the granulocytic leukocyte (1, 3-6). The binding of the proteins to the outer surface of their favorite cells, is expressed in terms of the augmentation of typical cellular functions and the increment of the cell's viability (6).

Studies on the leukophilic γ-globulin fraction, leukokinin, which comprise about 2-3% of the total serum γ-globulin has attracted major interest. First, leukokinin stimulates the phagocytic activity of polymorphonuclear leukocytes (PMNL). Phagocytosis is an event which constitutes a major line of defense for the host against the dangers of acute bacterial infections. Second, the whole activity of leukoninin could be attributed to a small daughter-peptide, derived from the protein and characterized as L-threonyl-L-lysyl-L-prolyl-L-arginine (6, 8-10). This basic tetrapeptide was named tuftsin, and it has been chemically synthesized (9, 11). A complete identity in physico-chemical and biological features of both the natural and synthetic compounds was established (9-11).

The revelation that tuftsin, is a natural and simple stimulator of phagocytosis has enormous, potential, clinical implications. It has generated rather extensive studies aimed mainly at: 1) ex-

Molecular and Cellular Biochemistry 41, 73-97 (1981). 0300-8177/81/0041-0073/$05.00.

ploring the scope of its functions *in vitro* and in vivo; 2) understanding the nature of its binding to phagocytic cells and 3) the mode of cell activation; 4) development of efficient synthetic methods; 5) understanding of structure-function relationships and 6) therapeutic evaluation.

In the following, we review in a concise manner the present major findings relevant to tuftsin. The reader is also referred to a number of general (6, 12, 13) and a more specific accounts (14) on this peptide.

The mode of tuftsin's generation from leukokinin ('biosynthesis')

It was shown by Najjar and his colleagues that tuftsin is generated from leukokinin during the process of the protein's attachment to the PMN-leukocyte and cell activation (12, 13). It is not clear yet whether the site of peptide production coincide with its locus of action. However, complete release from the parent protein molecule is probably needed in order to exert its functions. During its action, tuftsin is gradually degraded by serum enzymes and perhaps also consumed by the cells (13). The processed, [des-tuftsin]-leukokinin is rendered inactive and to perpetuate cell stimulation, the spent leukokinin is detached from the cell and replaced by a fresh molecule which serves as precursor for new tuftsin (9, 13, 15).

Two questions of relevance arise:
1) Is tuftsin an integral part of leukokinin?
2) How is leukokinin processed to yield tuftsin?

Examination of the sequence of the γ-globulin molecule (16) reveals that a tuftsin-moiety is located within the C_H2 domain of the Fc-fragment of the protein's heavy chain, in the vicinity of the carbohydrate binding site:

$$\underset{289 \qquad\qquad 292}{...\text{His-Asn-Ala-Lys-Thr-Lys-Pro-Arg-Glu-Gln-Gln-Tyr-Asx}...}$$

with CHO above Asx

Phagocytosis experiments (13–15), chemical analysis (15) and radioimmunoassay studies (17) on Fab and Fc fragments of IgG, confirm that only the latter protein portion can serve as a precursor for tuftsin. Further examination of the γ-globulin sequences (18), shows that tuftsin-moieties are present at location 289–292 in each of the IgG sub-classes (1→4). Minor alterations in close proximity to 'tuftsin' such as replacement of Tyr-296 by Phe or Glu-293 by Val do occur. In one example (MOPC-21; Mouse IgGl) Lys-290 is replaced by a Gln-residue. Does this modification have any functional implications in regard to tuftsin activity?

It was demonstrated that plasma membranes prepared from PMN-leukocytes of human blood contain an enzyme, leukokininase, capable of cleaving leukokinin, as does short treatment with trypsin, to yield tuftsin (10, 13). Leukokininase has been partially purified and characterized (10). Prior exposure of leukokinin to carboxypeptidase B action, followed by treatment with leukokininase, produced stochiometric amounts of arginine but not tuftsin (10, 15). It was thus concluded that leukokinin is a γ-globulin species incised at its Arg-292 Glu-293 peptide bond. The action of leukokinnase is hence, to specifically cleave the bond between Lys-288 and Thr-289, with subsequent release of tuftsin.

Leukokinin obtained from donors who underwent splenectomy or whose spleen functions are impaired, can still bind to PMNL and does contain a tuftsin-moiety (6). However, it does not stimulate phagocytosis by PMNL nor yield tuftsin upon treatment with leukokininase or trypsin (5, 14). Based on these findings, it was suggested that the spleen plays a crucial role in the initial step of leukokinin processing. It contains a specific enzyme, tuftsin endocarboxypeptidase, which splits the bond between Arg-292 and Glu-293 producing a protein, leukokinin-S (S for spleen), carrying free carboxyl terminal of tuftsin.

The existance of this splenic enzyme is based as yet only circumstantial evidence. Leukokinin-S, then circulates in the blood, finds its target cells, the PMNL, where it is finally processed by leukokininase to yield tuftsin. The two-step mechanism of tuftsin's generation from leukokinin, which is rather reminiscent of the production of angiotensin (12, 19) is schematically depicted in Fig. 1.

Metabolism of tuftsin

Phagocytic cells, the PMNL, the macrophage and the monocytes (see later section), contain specific receptors for tuftsin on their outer surface. It is not yet clear as to the fate of tuftsin following

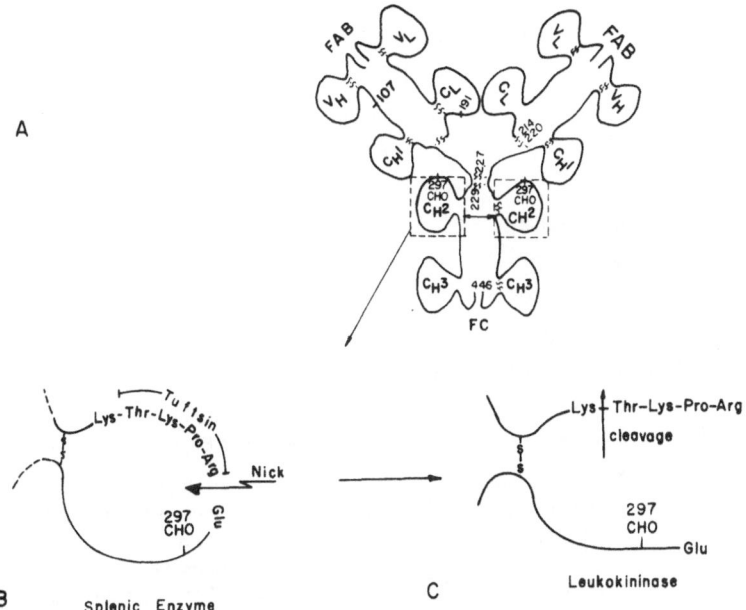

Fig. 1. The two step processing of leukokinin by splenic and leukocytic enzymes. (A) Leukokinin; (B) Cleavage by splenic enzyme yielding leukokinin-S; (C) Final splitting by leukokininase with release of tuftsin.

attachment to its target cells. It was shown, however, by Rauner *et al.* that the cytoplasm of rabbit PMN-granulocytes contain proline peptidases i.e., enzymes capable of degrading peptide bonds next to the C or N terminal of proline (20). Thus, tuftsin could be totally degraded yielding its constituent amino acids, Thr, Lys, Pro and Arg as well as the dipeptides Lys-Pro and Pro-Arg. If the peptide is internalized, following its attachement to cells these enzymes could serve as a mean of clearance. The lysosomal and the plasma membrane fractions of PMNL do not contain such enzymic activity (20).

The blood obviously contains a battery of various non-specific peptidases. It was not surprising to find, using an assay of phagocytosis, that indeed tuftsin is degraded by blood-enzymes though at a rather moderate rate (21). If the future use of tuftsin as a drug is to be considered, its stabilization toward enzymatic degradation is of utmost importance. In an attempt to prepare stable long-acting analogs of tuftsin, Stabinsky *et al.* (22) have synthesized a series of derivatives in which the N-terminal threonyl residue of tuftsin is cyclized and hence more stable to enzymic degradation. Preliminary results indicate that [O = C◇Thr¹] tuftsin, though possessing only 50–60% of tuftsins activity, fulfills such requirements (22).

Chemical synthesis of tuftsin

Ever since the isolation and structural elucidation of tuftsin by Nishioka *et al.* (8–10) it has been chemically synthesized in many laboratories. Although the preparation of a tetrapeptide by present day methodology of peptide synthesis seems an easy task, the synthesis of tuftsin may create certain difficulties. Three of the constituent amino acids, threonine, lysine and arginine, are tri-functional and need, depending on the procedure employed, adequate protecting groups. The fourth amino acid, proline, is a notorious and problematical amino acid from the synthetic point of view. However, using appropriate precautions, tuftsin has been synthesized in practically all the available synthetic approaches, e.g.: 1) conventional preparation in homogeneous solution phase (22–31); 2) solid-phase synthesis (9, 11, 32); 3) polymeric-reagents synthesis (25, 27). In these syntheses a large variety of protecting groups, for various amino acid functions, and of coupling agents, for peptide bond formation, were used. The different tuftsin preparation synthesized were quite similar to each other in their physico-chemical and biological characteristics. It is pertinent to note here, that very, highly pure tuftsin is crucial in order to obtain optimal biological activities. Moreover, various synthetic approaches do yield impurities

76

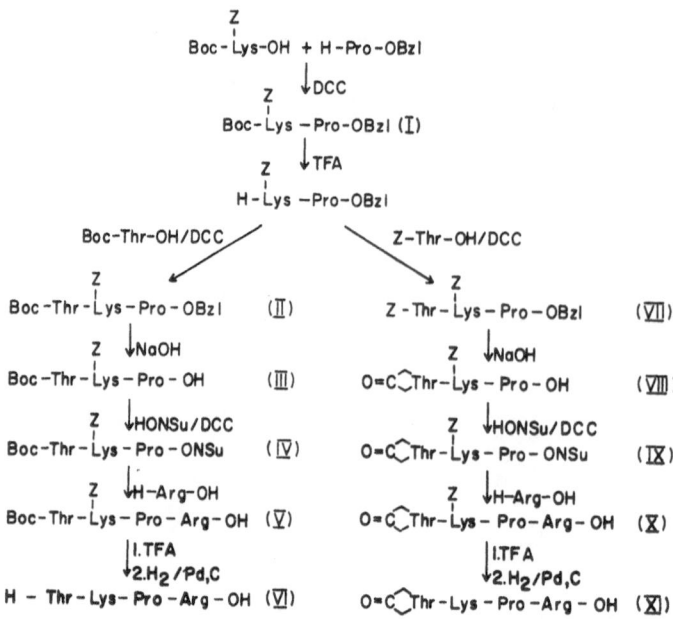

Fig. 2. Synthesis of tuftsin and [O = C↺Thr¹]tuftsin. Replacement of arginine by [³H]arginine or by various basic amino acids at synthetic stage IV→V yielded [³H]tuftsin (46) or 4-position analogs (22, 33).

which are closely related to tuftsin. They are sometimes detectable with difficulty and may inhibit tuftsin's action.

It is not intended here to detail all the synthetic procedures used in the preparation of tuftsin and the reader is referred to the relevant above mentioned literature. However, one of the conventional routes used in our larboratory to synthesize tuftsin, some of its analogs and [³H]tuftsin is schematically illustrated in Fig. 2.

Structure-function relations in tuftsin

Aiming, to shed light upon the mode of tuftsin's action on phagocytic cells and its specificity, over 60 related analogs has been synthesized in various laboratories and *via* different synthetic approaches (e.g., Fig. 2). The peptides prepared and their biological potencies, relative to that of tuftsin, are listed in Table 1. The data summarized clearly indicate that the integrity of the entire sequence of tuftsin is crucial for the expression of its maximal biological activity. Structural alterations, often very minor, drastically effect activity and may lead to a marked decrease, abolishment, or even inhibition of action. Thus, activity was found to be significantly impaired by: (a) amino acid substitution or deletion at the N-terminal, C-terminal or within the chain; (b) absence of free amino and

carboxyl functions at the peptide termini; (c) substitution of the guanidine side chain of the arginine residue which lead to neutralization or partial masking of its positive charge or to the altering of its distance from the peptide backbone; and (d) lengthening of the peptide chain by addition of amino acids at either C or N-terminus. Certain modification seem, however, to be quite 'permissible' and in particular the interchange of Lys² and Arg⁴ or substitution of these basic amino acid by each other (28).

It seems that four structural 'working rules' should be considered if a significant biological activity is to be anticipated from a tuftsin analog. First, the over-all positive charge and its distribution within the peptide should be kept as close as possible to that of tuftsin. Second, a proline residue must be implanted in close proximity and ideally between the two positive charges. Third, the integrity of the N-terminal threonine residue should be maintained and fourth, chemical modifications such as chain extension will only be at the C-terminal of tuftsin.

Some discrepancies exist between the biological activities of certain identical analogs prepared and tested in different laboratories by, basically similar assays. This is particularly evident in the case of retro-tuftsin, Arg-Pro-Lys-Thr, which showed either the same activity as tuftsin (37) or no activity at all (28, 38, 39). It is recommended, usually to

Table. 1. Relative augmentory and inhibitory activity of tuftsin analogs as compared with tuftsin.

No.	Peptide analog	Relative activity	Inhibitory activity
1	Thr-Lys-Pro-Arg	1.0	—
2	Thr-Lys-Pro-Homoarg	0.95(P)[33]; 1.0(R)[33]	N.D.
3	Thr-Lys-Pro-Lys	0.70(P)[33]; 0.65(R)[33]	N.D.
4	Thr-Lys-Pro-Arg-NH$_2$	0.60(P)[33]; 0.65(R)[33]	N.D.
5	Thr-Lys-Pro-N^G-methyl-Homoarg	0.50(P)[33]; 0.65(R)[33]	N.D.
6	Thr-Lys-Pro-Norarg	0.50(P)[33]; 0.65(R)[33]	N.D.
7	Thr-Lys-Pro-N^G-methyl-Arg	0.45(P)[33]; 0.35(R)[33]	N.D.
8	Thr-Lys-Pro-Ala	0.65(P)[29, 34]	N.D.
9	Thr-Lys-Pro-Gly	0.45(P)[34]	N.D.
10	Thr-Lys-Pro-Arg-Gly	0.42(P)[33]; 0.35(R)[33]	N.D.
11	Thr-Lys-Pro-D-Arg	0.35(P)[33]; <0.2(R)[33]	N.D.
12	Thr-Lys-Pro-His	0.34(P)[33]; 0.35(R)[33]	N.D.
13	Thr-Lys-Pro-Arg (NO$_2$)	0(P)[27]; 0(R)[27]	0(P)[27]; 0(R)[27]
14	Thr-Lys-Pro	<0.1(P)[33]; <0.2(R)[33]	N.D.
15	Thr-Lys-Ala-Arg	0.60(P)[35]	N.D.
16	Thr-Lys-Gly-Arg	0.35(P)[33]; <0.2(R)[33] 0(P)[35]	N.D.
17	Thr-Lys-Thr-Arg	0.8(P)[34, 35]	N.D.
18	Thr-Lys-Leu-Arg	0.5(P)[34, 35]; 0.2(P)[36]	N.D.
19	Thr-Lys-ILeu-Arg	0.4(P)[34, 35]	N.D.
20	Thr-Lys-Val-Arg	0.4(P)[29, 35]	N.D.
21	Thr-Lys-Sar-Arg	0.1(P)[35]	N.D.
22	Thr-Lys-Phe-Arg	0.2(P)[34, 35]	N.D.
23	Thr-Lys-Tyr-Arg	0.35(P)[34, 35]	N.D.
24	Thr-Lys-His-Arg	0.15–0.2(P)[34, 35]	N.D.
25	Thr-Lys-Lys-Arg	0.45(P)[35]	N.D.
26	Thr-Lys-Ser-Arg	0–0.1(P)[34, 35]	N.D.
27	Thr-Lys-Asp-Arg	0.15(P)[35]	N.D.
28	Thr-Lys-Gln-Arg	0.45(P)[35]	N.D.
29	Thr-Lys —— Arg	0.25(P)[33]; 0.35(R)[33]	N.D.
30	Thr-Lys-Arg-Arg	0.85(P)[35]	N.D.
31	Thr-Arg-Pro-Arg	1.0(P)[28, 37, 38, 39]	N.D.
32	Thr-Orn-Pro-Arg	0.45(P)[34, 37, 40]	N.D.
33	Thr-Leu-Pro-Arg	0.25(P)[34]	N.D.
34	Lys-Lys-Pro-Arg	0.30(P)[27]; <0.3(R)[27]	<0.3(R)[27]
35	Val-Lys-Pro-Arg	0(P)[27]; <0.3(R)[27]	0.30(P)[27]; <0.3(R)[27]
36	Leu-Lys-Pro-Arg	>1.0(P)[41]	N.D.
37	Ser-Lys-Pro-Arg	0.20(P)[27]; <0.3(R)[27]	0.78(P)[27]; <0.3(R)[27]
38	Ala-Lys-Pro-Arg	0(P)[27, 34]	0.65(P)[27]; 0.65(R)[27]
39	Tyr-Lys-Pro-Arg	N.D.(P)[27]; <0.3(R)[27]	N.D.(P)[27]; <0.3(R)[27]
40	0 = C‹›Thr-Lys-Pro-Arg	0.54(P)[33]; 0.65(R)[33]	N.D.
41	Acetyl-Thr-Lys-Pro-Arg	0(P)[27]; <0.3(R)[27]	0.44(P)[27]; <0.3(R)[27]
42	PAPA-Thr-Lys-Pro-Arg	0(P)[27]; <0.3(R)[27]	0.22(P)[27]; <0.3(R)[27]
43	Lys-Thr-Lys-Pro-Arg	0(P)[34]	N.D.
44	Lys-Pro-Arg	0(P)[27]; <0.3(R)[27]	0.44(P)[27]; 0.65(R)[27]
45	0 = C⌣Thr-Lys-Pro-Homoarg	0.45(P)[33]; 0.65(R)[33]	N.D.
46	0 = C⌣Thr-Lys-Pro-Lys	0.20(P)[33]; 0.35(R)[33]	N.D.
47	0 = C⌣Thr-Lys-Pro-Norarg	0.15(P)[33]; <0.2(R)[33]	N.D.
48	0 = C⌣Thr-Lys-Pro-His	0.10(P)[33]; 0.35(R)[33]	N.D.
49	0 = C⌣Thr-Lys-Pro-Arg-NH$_2$	0.30(P)[33]; <0.2(R)[33]	N.D.

Table 1 (continued)

No.	Peptide analog	Relative activity	Inhibitory activity
50	Thr-Pro-Lys-Arg	N.R.(P)[42]	N.D.
51	Thr-Lys-Arg-Pro	N.R.(P)[42]	N.D.
52	Thr-Arg-Pro-Lys	1.0(P)[28, 38, 39]; N.R.(P)[37]	N.D.
53	Arg-Pro-Lys-Thr	1.0(P)[37];0(P)[28, 38, 39]	N.D.
54	Lys-Thr-Arg-Pro	N.R.(P)[42]	N.D.
55	Lys-Thr-Pro-Arg	N.R.(P)[42]	N.D.
56	Thr-Lys-Lys-Ala	0.5(P)[34]	N.D.
57	Thr-Lys-Ala-Ala	0.13(P)[29, 34]	N.D.
58	Thr-Gly-Gly-Lys	0-0.1(P)[29, 34]	N.D.
59	Thr-Ala-Arg-Lys	0.70(P)[29, 34]	N.D.
60	Thr-Ala-Val-Arg	0.75(P)[29, 34]	N.D.
61	Thr-Orn-Pro-Ala	0.3(P)[40]	N.D.
62	Thr-Pro-Lys-Ala	0.1(P)[40]	N.D.
63	Thr-Lys-Pro-Pro-Arg	O(P)[9, 10, 11]; N.R.(P)[42]	+(P)[9, 11]
64	Thr-Gly-Lys-Pro-Arg	N.R.(P)[37]	N.D.
65	Lys-Pro-Pro-Arg	O(P)[13]	+(P)[13]
66	(Lys)$_{(3, 4, 6, 20)}$	O(R)[27]	O(R)[27]
67	Leu-Lys-Lys-Ala	0.5(P)[40]	N.D.

A value of *1.0* is taken as the activity of tuftsin; N.D., not determined; P, phagocytosis assay; R, NBT-reduction assay – formazane extracted from cells with pyridine and determined at wavelength of 515 nm; N.R., activity not reported; +, strong inhibition; PAPA, p-aminophenylacetyl.

solve such controversies by analyzing the analogs simultaneously. It is possible, however, that the non-active compound do contain some trace amounts of a potent peptide inhibitor.

Table 1 points at several synthetic inhibitors of tuftsin's different biological activities. Thus, Thr-Lys-Pro-Pro-Arg (9, 11), Ser-Lys-Pro-Arg (27), Ala-Lys-Pro-Arg (27), Lys-Pro-Arg (27), and to lesser extent other peptides, can block the stimulation of phagocytosis induced by tuftsin on phagocytic cells. The first of these analogs proved also to inhibit augmentation of the motility of PMN-leukocytes by tuftsin (11). Several analogs like Ala-Lys-Pro-Arg or Lys-Pro-Arg can inhibit the stimulatory effect of tuftsin on the reduction of the dye nitroblue tetrazolium by PMN-leukocytes (27). All of the inhibitory influences, which are powerful means for studying the mechanism of tuftsin's action, were exerted in *in vitro* assays. It would be of utmost importance to examine whether similar effects do exist in *in vivo* systems.

It is notable that, only one, [Leu¹]tuftsin, of the available synthetic analogs was reported to exhibit a higher biological potency than that of tuftsin (41).

However, no information about the relative activity of this peptide is given.

The rather strict architectural requirements for biological activity of tuftsin may suggest that in order to exert its full stimulatory effects, the peptide assumes a particular three dimensional structure. Does tuftsin indeed have a well defined conformation under physiological conditions in aqueous solution, or perhaps it acquires one, upon binding to its specific receptors on target cells?

On the structure of tuftsin

The possibility that tuftsin may assume a preferred conformation in solution was investigated. Although only a limited number of relevant reports have been published, it seems that rather conflicting conclusions are drawn. Konopinska and colleagues have suggested from theoretical considerations based on X-ray, N.M.R., and I.R. data (42a) as well as recent I.R. studies (34) that tuftsin has a tendency to form a $4 \rightarrow 1$ hydrogen bonded β-turn. In addition, this structure, it is claimed, is further stabilized by ionic interaction between the carboxyl terminal of Arg[4] and the α-amino terminal of Thr[1]. Based on circular dichroism study with tuftsin and tuftsin esterified at the C-terminal, it was suggested by Vicar et al. (26), that the peptide may have some pseudo-cyclic structure in which the negative carboxylic group of Arg[4] ionically interacts with the positively charged ammonium side chain of Lys[2]. Scheraga and co-workers have studied the structure of tuftsin by conformation energy-minimization calculations (43). It was concluded that the peptide may assume numerous low-energy conformations. These various conformations share, however, a general characteristic described as a 'hairpin with two split ends' where each of the splits is composed of one terminus of the backbone and a Lys or Arg side chain (43). This parental-structure, is in line with results obtained from a study on the phagocytosis stimulatory activity of some tuftsin's analogs (29). Both studies suggest that the positive charge on the Lys[2] and Arg[4] residues have only a minor influence upon biological activity (29, 43), whereas the Thr[1] residue is not essential at all for activity (43). These last conclusions entirely oppose our findings that not only the integrity, but also the exact spatial-location and the complete accessibility of the two side-chain positive charges are crucial for expressing maximal biological capacity (33). Moreover, the presence of an intact Thr[1]-residue in tuftsin is indispensable, if a high biological potency is sought (27, 31, 33). Najjar and coworkers (44) have recently studied the solution conformation of tuftsin by both carbon-13 and proton nuclear magnetic resonance and concluded that there is no evidence for a normal 4→1 β-turn and in fact any other preferred structure in aqueous solution. Similar conclusions were obtained in an ^{1}H and [13]C N.M.R. study on tuftsin and Thr-Lys(Cbz)-Pro-Arg carried out in D_2O and CD_3OD by Sekacis et al. (45). Measurements done in DMSO indicate, however, that tuftsin may assume a β-turn structure in this solvent. Our own structure-function studies (33) on tuftsin analogs modified at the peptide's C, or N-terminals, or both, support the notion that tuftsin has a rather random structure in water. The high specificity of its biological action as well as its binding to specific receptors (see below), are therefore indicative of the fact that tuftsin may assume an 'active conformation' only upon adhering to its target cells.

Specific binding sites for tuftsin

Chemically induced alterations of the tuftsin molecule indicate that a specific structural integrity must be maintained for the expression of this peptide's full biological activity (see section on structure-function relations). These results have led to studies on tuftsin's receptor site on various cell types.

Periphary cells

A synthetic route for the incorporation of tritiated arginine into tuftsin has been devised by Stabinsky et al. (46). During synthesis, the chemical composition of the tuftsin molecule is not altered. This is accomplished by coupling the activated and protected tripeptide Boc-Thr-Lys-(Cbz)-Pro-OSu to tritiated arginine. Subsequent deprotection and purification yield a highly purified, radiolabelled (specific activity 9-24 Ci/mmol), tuftsin molecule, indistinguishable in its chemical and biological properties from unlabelled synthetic tuftsin (Fig. 2).

Using [3H]tuftsin, Stabinsky et al. (46) performed binding studies at 22 °C on various cell types found in the blood circulatory system. It was clearly demonstrated that specific receptor sites exist on PMNL-cells and that this binding is a time dependent reversible, process. Saturation of sites as displayed in Fig. 3A reaches a level of 280 nM and Scatchard analysis of this data (Fig. 3B), indicates that only one type of binding site exists for tuftsin. The equilibrium dissociation constant, Kd, is 1.3×10^{-7} M, close to the concentration of tuftsin which elicits a half maximal response in human PMNL. Assuming an equimolar ligand-receptor complex, 50 000 sites are present on each PMNL-cell.

The specificity of PMNL binding sites for [3H]tuftsin was further established by its capacity to competitively bind with unlabelled tuftsin as well as other structurally related analogs (46). Fig. 4 displays the results of these assays. When the concentration of unlabelled to labelled tuftsin is one to one, inhibition of binding is about 50%, a critical indication of tuftsin's specificity. Two biologically inactive analogs Des-Arg[4]-tuftsin and N^G-NO_2-Arg[4]-tuftsin were unable to compete with tritiated tuftsin even at very high concentrations. On the other hand, [Homoarg[4]]tuftsin, an analog with

80

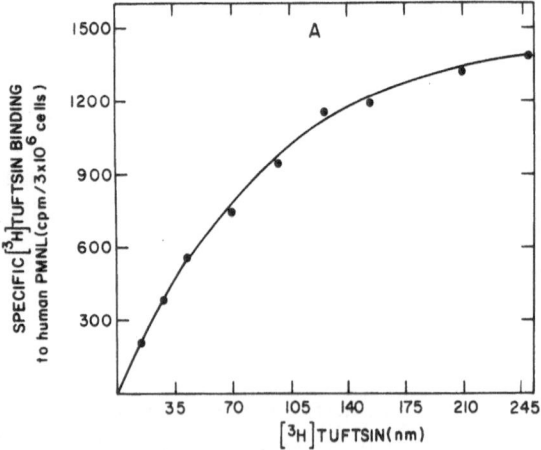

Fig. 3.A Specific binding of [³H]tuftsin to human PMNL as a function of concentration fo [³H]tuftsin.

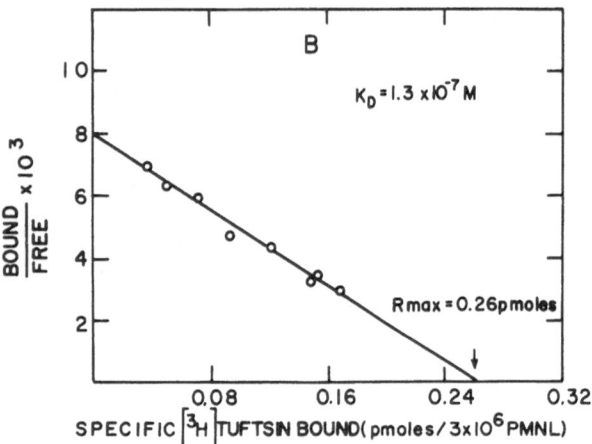

Fig. 3B. Scatchard analysis of binding data (46).

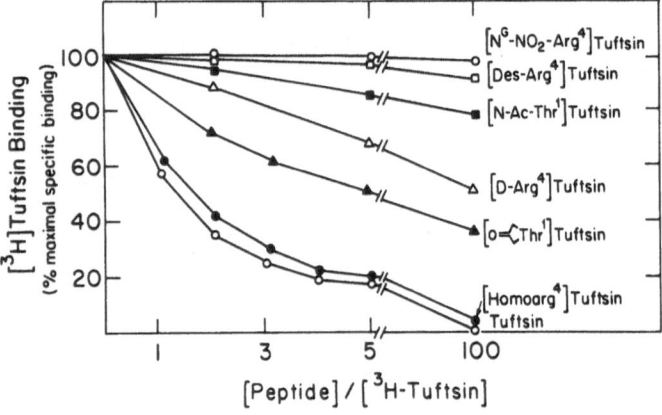

Fig. 4. Effect of unlabelled tuftsin and its analogs on the binding of [³H]tuftsin to human PMNL (46).

tuftsin like activity, competes for tuftsin binding sites to the same extent as unlabelled tuftsin. An analog which stimulates phagocytosis by 50–60%, [0=C͡Thr¹]tuftsin, binds rather well, although not to the same degree as tuftsin. Stabinsky *et al.* have also prepared tritium-labelled N-acetyl-tuftsin which does not demonstrate any capacity to bind to tuftsin receptor. These analogs as well as others, indicated in Fig. 4, deliniate a direct correlation between the potency of the phagocytic response and the degree of specific binding.

Several other cell types were tested for their ability to interact with tuftsin *via* specific receptor sites. Purified lymphocyte preparations were found to bind less than 5% of tritiated tuftsin as compared with PMNL cells. This threshold binding capacity has been attributed either to low affinity of interaction or the complete absence of specific sites for tuftsin (46).

Mononuclear preparations (79% lymphocytes and 21% monocytes) showed a specific binding capacity of 41% of that calculated for PMNL. This binding capacity was therefore attributed mainly to the monocytes present in the mixture. Saturability of tuftsin binding sites is very close to that of PMNL, with a K_d, equal to 1.25×10^{-7} M, and contains approximately 100 000 tuftsin sites per cell. No specific binding was detected on erythrocytes (46).

A second approach, employed by Fudenburg and colleagues (47), was the chemical introduction of a radiolabelled tracer into the C-terminal portion of tuftsin. In this manner tuftsin-¹⁴C-methyl ester and ¹²⁵l-tuftsyl-tyrosine were prepared and used for binding studies on circulatory cells.

Nair et al. utilizing these radiolabelled tuftsin-like materials, showed that specific sites for tuftsin are present on PMNL (47). These findings are supported by binding competition experiments using unlabelled synthetic tuftsin, demonstrating specificity as well as reversibility. Preincubation with chicken antisera specific to tuftsin, abolishes all binding activity. Incubation of these labelled compounds with cells displayed a temperature dependency, where optimum binding was achieved at either 4 °C with or without prior incubation at 37 °C. Binding carried out at 22 °C was minimal. Radiolabelling of tuftsin by modification at the N-terminal led to loss of binding ability to tuftsin's receptor sites.

Similar assays using a pure lymphocyte preparation display a lower precentage of binding in comparison to PMNL cells (66%). However, only binding displayed by ^{125}I-tuftsyl-tyrosine could be considered statistically significant (47). Monocytes, on the other hand, bind well with both radio-labelled molecules (47).

In general, the results obtained from both approaches indicate that specific site exist on PMNL and monocytes for tuftsin. Certain analogs with various modifications, especially at the N-terminal have no ability to bind specifically to these receptors. Major differences, however, are apparent, upon critical examination and comparison of both approaches. The first of these discrepancies is the temperature dependence displayed in the assays conducted by Nair *et al.* where minimal binding was found at $22\,^{\circ}$C, the temperature at which Stabinsky *et al.* demonstrated significant binding of tuftsin. An even more serious discrepancy is the appreciable binding of tuftsin to lymphocytes as shown by Nair and collaborators, in contrast to the minimal binding indicated by Stabinsky and coworkers. From the latter results, Nair et al. have reasoned that tuftsin activity is mediated by specific binding to the Fc receptor. This conclusion should be considered premature due to the above descrepancy in which a 'native', intact, tuftsin molecule shows little or no binding to lymphocytes while only a greatly modified tuftsin material could display significant binding.

Macrophages

Demonstration of tuftsin binding sites on monocytes, naturally led to the search for such receptor sites on macrophages. Bar-Shavit *et al.* studied the effect and binding of tuftsin on normal, as well as a variety of *in vivo* stimulated mouse peritoneal macrophage populations (48). The binding of ³H-tuftsin to thioglycollate-stimulated macrophages qualitatively mimics those findings found for monocytes and PMNL. Binding was shown to be rapid, reversible and saturable (see Fig. 5) with a K_D equal to approximately 5.3×10^{-8} M as calculated from the Scatchard plot (see Fig. 5). Only one population of sites exist and the number of sites calculated per macrophage is about 72 000.

Studies performed using unlabelled tuftsin as well as some analogs, indicate again that the magnitude of the binding is indicative of the phagocytic response. Thus, the two biologically inactive analogs Des-Arg⁴-tuftsin and N-acetyl-tuftsin showed no competition fo these binding sites. On the other hand, D-Arg⁴-tuftsin, an analog which can stimulate phagocytosis to a small degree, exhibits a corresponding capacity for binding (48).

No significant differences could be observed for the various stimulated macrophage populations. It is of interest that normal macrophages do not display any substantial differences in tuftsin binding in comparison to its activated counterpart (48).

Lymphocyte suspensions of thymic or splenic origin showed no capacity to bind tritiated tuftsin. This is also true of tertiary mouse embryo fibroblast monolayers (48).

In vitro differentiated cells

Two additional types of cells were tested for the presence of functional tuftsin binding sites (49). Mouse bone marrow cells differentiated *in vitro* into mononuclear phagocytes exhibits such tuftsin

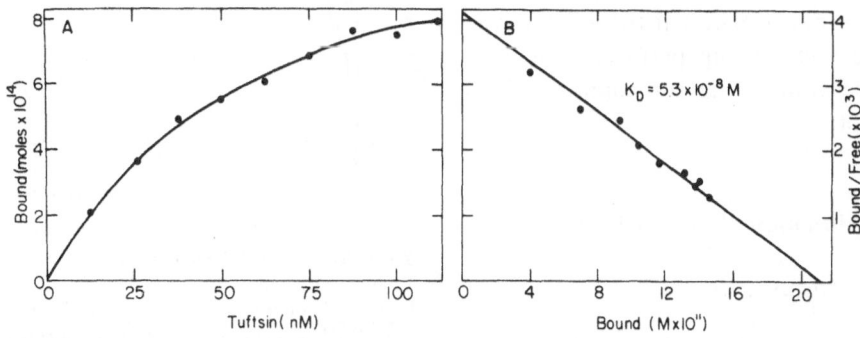

Fig. 5. Specific binding of [³H]tuftsin to mouse peritoneal thioglycollate stimulated macrophges. Left, concentration dependence of binding. Right, Scatchard analysis of binding data (48).

binding sites and their phagocytic response is augmented by tuftsin. The binding capacity of these cells was shown to be higher by a factor of three as compared to *in vivo* stimulated or normal macrophages (based on mg protein). These findings are independent of variations in experimental conditions. Removal of condition media results in the cessation of DNA synthesis and of cell proliferation but had little or no effect on the binding capacity of tritiated tuftsin to these cells. This indicates that the expression of the tuftsin receptor is independent of cell cycle or proliferation (49).

The second cell type tested, which supports the above findings, was the lymphoma derived macrophage-like line called P388D1 (49). This cell line displays a capacity to bind tuftsin equal to that of peritoneal macrophage and is stimulated to phagocytosis by tuftsin. The existence of such specific binding sites coupled with the fact that Fc and C3b receptors exist (50) lends credence to the possibility that this cell line is of mononuclear phagocyte origin. In general, it appears that the tuftsin receptor exists on all phagocytic type cells independent of origin.

Defective PMN-leukocytes receptor?

The study of Constantopoulos *et al.* (51), on the phagocytic activity of PMNL-cell taken from various leukemia patients, is perhaps of relevance here. Cells obtained from seven patients suffering from acute granulocytic leukemia and from six patients with myelofibrosis, whether in relapse or remission, proved normal or near-normal basal phagocytic capacity. The cells failed, however, to show any stimulation of phagocytosis, either by addition of autologus serum or by tuftsin added at doses even as high as 1000-fold greater than optimal. A study on the binding of [³H]tuftsin to the defective phagocytes is currently being performed in our laboratory, aimed at detecting and characterising their tuftsin receptors.

The scope of tuftsin's biological functions

Enhancement of phagocytosis

Phagocytosis, the process of envelopment and digestion of bacteria or other foreign bodies, is undoubtedly the most obvious and prominent function of phagocytic cell, the granulocytic leukocyte and the macrophage. Naturally, stimulation of their capacity to phagocytize was the first assay of choice in evaluating the role of leukokinin, i.e., tuftsin's precursor, and thereafter tuftsin and its synthetic analogs. Cells obtained from different mammalian donors were studied: e.g. PMN-leukocytes of dog (9, 11), guinea pig (9, 29, 39, 52), rabbit (52, 53), cow (54) and human (9, 27, 33, 51); macrophages from mouse peritoneum (30, 48, 52), spleen and liver (30) and from rabbit lung (52). Assays were performed in cell suspensions, e.g. (9, 51), or in monolayers, e.g. (27, 48), according to the expertise of the research laboratory. Four types of particulate matters were primarily employed: bacteria such as staphylococci (9–11, 29, 54); yeast cells like *Saccharomyces oviformis* (27, 46, 48, 55), IgG-coated sheep red blood cells (IgG-coated SRBC) (48) and ⁵¹Cr-IgG-coated-SRBC (55).

Using a variety of these assays it was found that tuftsin augments the phagocytosis capacity of both polymorphonuclear leukocytes and macrophages. The sensitivity of both these cell types to tuftsin is quite similar and is displayed in the dose-response curves for PMN-leukocytes (Fig. 6) and for macrophages (Fig. 7). Maximal stimulations, usually double or even triple the level attained by

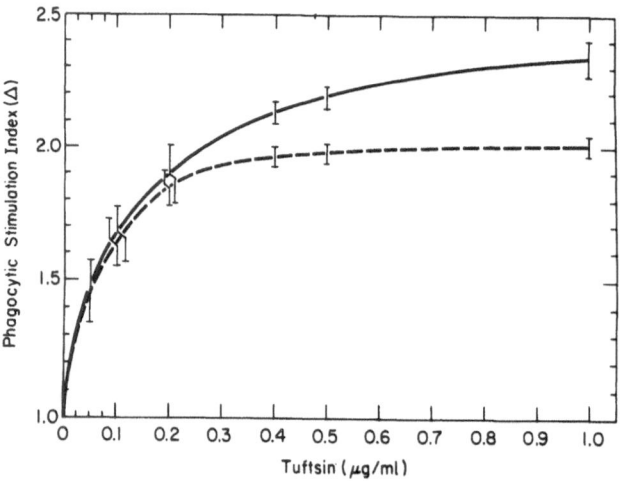

Fig. 6. The effect of tuftsin on the phagocytosis of heat-killed yeasts by human PMN-leukocytes (27).

Phagocytosis index $\Delta =$

$\dfrac{\% \text{ of PMN's containing yeasts with tuftsin}}{\% \text{ of PMN's containing yeasts without tuftsin}}$ (--)

$\dfrac{\text{number of yeasts ingested by 400 PMN's with tuftsin}}{\text{number of yeasts ingested by 400 PMN's without tuftsin}}$ (-)

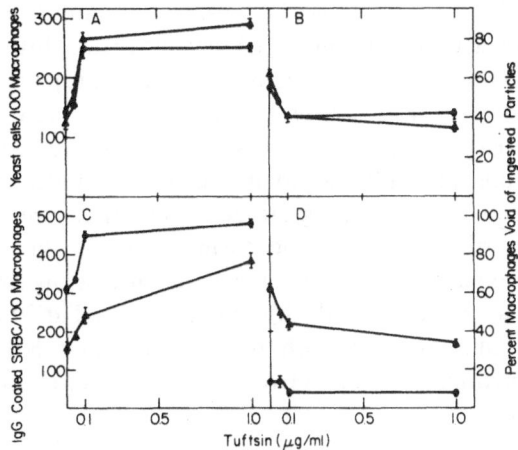

Fig. 7. Effect of tuftsin on the phagocytosis of (●—●) normal, (▲—▲) thioglycollate stimulated macrophages. A and B represent yeast cell phagocytosis whereas C and D represent IgG-coated SRBC phagocytosis (48).

non-stimulated cells, are observed at peptide concentrations of 100–200 ng/ml (200–400 nM). Rather significant values are noticeable even at hormonal-like levels of 5–10 ng/ml. Many synthetic analogs of tuftsin have been evaluated using assays of phagocytosis. The dramatic dependence of activity upon peptide's structure has already been discussed and the reader is referred to earlier paragraphs and to Table 1.

The fact that the stimulatory response of the phagocytic cells to tuftsin was practically independant of the donor, of the different sources within same species, and of the mode of sensitization employed to elicit cell recruiting and response, strongly indicates a common mechanism of excitation. The existance of specific receptors for tuftsin on surfaces of phagocytes has been described. Since tuftsin originates from the Fc-portion of the IgG molecule, and the participation of the Fc moiety is well documented in the phagocytosis process, speculation has arisen as to whether the effect of tuftsin is exerted through pertrubation of Fc-receptors. As illustrated in Fig. 7 it seems that tuftsin's augmentation of the phagocytic response is mediated both *via* Fc-receptors and via non-specific receptors (48). However, one should not exclude the possibility that tuftsin indeed binds only to Fc-receptors thus causing a massive membrane perturbation which can trigger phagocytosis *via* non-specific receptors as well.

Increasing the respiratory burst of phagocytes

The term respiratory-burst applies to a chain of redox metabolic changes, of a complex nature, occurring within the phagocytic cell upon its exposure to appropriate excitation (for review see Babior (56)). A direct correlation exists between these intracellular events, enhancement of the cell's bactricidal activity and increased phagocytic response (56, 57). An increase in the respiratory-burst of phagocytic cells is apparent upon exposure to tuftsin. This provides an additional measure of the peptide's activity and is indicative of its bactericidal capacity. The hexose monophosphate shunt (HMPS) is an example of a relevant intracellular oxidatory-event. It is a metabolic pathway in which glucose, a hexose, is oxidized to yield stoichiometric amounts of pentose and carbon dioxide, while $NADP^+$ acts as the oxidizing agent (i.e., electron acceptor). The pace of $NADP^+$ production from oxidation of NADPH determines shunt activity, and it is known to be enhanced up to 10 folds during phagocytosis (56, 57). As a result of this activation, oxygen is reduced to hydrogen peroxide which along with other highly reactive intermediates, i.e. superoxide ion and hydroxyl radical, participate in the mechanism of bacterial killing.

Several analytical assays are available for monitoring HMPS-activity (56, 57), two of these were utilized in our laboratory:

1. Nitroblue tetrazolium (NBT) reduction

Stimulated PMN-leukocytes are capable of reducing the yellow water-soluble compound NBT, converting it into insoluble blue formazan. The electron-acceptor dye initially enters the cell, probably *via* phagocytosis and-or pinocytosis, where it is reduced by a superoxide anion (O_2^-) generated during the respiratory-burst (56, 57, 58). The NBT-test has a useful clinical value in diagnosis of redox capabilities of PMN-leukocytes in patients with recurrent bacterial infections (59).

Tuftsin was found to exhibit a considerable stimulatory effect on NBT-reduction by PMNL, similar to that of endotoxin (B_4, lipopolysaccharide), a known activator of HMPS (60). Many synthetic tuftsin analogs were also screened in the NBT-test (60, 27, 33). Some of these did metabolically activate PMNL cells, though as a rule, with the exception of [Homoarg⁴]tuftsin, always to a much lesser extent than tuftsin. Several were inactive, while [Ala¹]tuftsin and [Des-Thr¹]tuftsin inhibited the action of tuftsin (Table 1). It is worth mentioning that both inhibitory peptides block the action of endotoxin, a finding which may suggest that the cellular binding sites for the lipopolysaccharide and for tuftsin share common determinants or at least have some topographical proximity. It is also notable, as found by Gottlieb (61), that of the four phosphocellulose immunoglobulin fractions, only leukokinin could stimulate NBT-reduction by PMN-leukocytes.

The stimulatory effect of tuftsin on NBT-reduction by phagocytes was also studied by Iguchi and Nakazawa (62).

$^{14}CO_2$ release from [^{14}C-1-glucose].

As previously described, oxidation of glucose *via* the HMPS yield CO_2 gas whose origin is carbon-1 of the hexose (56, 57). Metabolism of [^{14}C-1-glucose] yields $^{14}CO_2$ which is easily utilized in monitoring shunt activity. Stabinsky (63) has shown that tuftsin as well as several of its analogs, can stimulate glucose oxidation in PMN-leukocytes from human or guinea pig origin (Table 2). As shown in Table 2 both resting and phagocytozing cells responded to various peptide analogs. It seems possible, however, that stimulation of the former cells is due to increment in pinocytosis. It is rather interesting, and yet unclear, that [Ala¹]tuftsin and [Des-Thr¹]tuftsin, inhibitors of both phagocytosis and NBT-reduction, proved inactive in the glucose oxidation assay.

Immunostimulation by tuftsin

Recent studies have strongly suggested that the activation by antigen of T-lymphocytes, resulting in antibody production, depends on the presence of macrophages (64–67). It appears that a macrophage-antigen complex triggers a sequence of

Table 2. The effect of tuftsin and some analogs on [^{14}C-1]glucose oxidation by resting and phagocytosing polymorphonuclear leukocytes (63).

Compound (3 × 10⁻⁷ M)	Stimulation (%) Normal Human PMNL[a]		Stimulation (%) Guinea pig PMNL[b]	
	Resting cells	Phagocytosing[c] cells	Resting cells	Phagocytosing[a] cells
Tuftsin	30	52	27	45
[Homoarg⁴]tuftsin	26	44	25	40
[O = C∿Thr¹]tuftsin	12	29	15	22
[Arg-NH₂⁴]tuftsin	10	22	12	28
[D-Arg⁴]tuftsin	6	8	7	10
[Ala¹]tuftsin	–	–	–	5
[Des-Thr¹]tuftsin	–	–	–	–

Stimulation values are calculated according to:

$$\{[CPM_{(peptide)} - CPM_{(blank)}]/[CPM_{(blank)}]\} \times 100$$

CPM = $^{14}CO_2$ cpm per 1 μCi [^{14}C-1]glucose per 10 min per 10⁷ cells.

CPM (blank) values were 9320 ± 1869 and 15 750 ± 2340 for resting and phagocytosing normal human PMNL respectively; the corresponding values for guinea pig peritoneal exudate PMNL were 20 950 ± 3450 and 39 120 ± 6080 respectively.

CPM (tuftsin) values were 12 200 ± 2130 and 23 680 ± 3240 for resting and phagocytosing human PMNL respectively; the corresponding CPM (tuftsin) values for guinea pig peritoneal PMNL were 26 620 ± 4110 and 56 550 ± 11 130 respectively.

[a] Peripheral PMNL.

[b] Caseinate-stimulated peritoneal PMNL.

[c] Particles used: unopsonized lattex particles (0.81 μ) 0.3 mg/10⁷ PMNL.

intercellular interactions eventually leading in the production of effector T-lymphocytes. Yet, various aspects of the mechanism of this antigen-specific macrophage-dependent 'education' of T-cells are largely unknown. One of the open questions of relevance is for example, whether a specific stimulus or signal is required for the initial formation of the macrophage-antigen conjugate.

Knowing that macrophages carry specific surface-receptors for tuftsin, and that their perturbation by this peptide invokes various cellular functions, Tzehoval *et al.* (68–70), were inspired to study the possibility that tuftsin augments the cells capacity to 'educate' T-lymphocytes. Thus, monolayers of macrophages were incubated in the absence or presence of various concentrations of tuftsin and a constant amounts of antigen (Keyhole limpet hemocyanin, KLH) for 4 h. Spleen cells were then seeded on the macrophage monolayers, cultured for 24 h, removed and irradiated to prevent replication, and injected into the footpad of syngeneic mice. Six days later the regional popliteal lymph node cells were removed and tested for specific reactivity to the antigen by the host-recruited lymphocytes, assayed by ^{3}H-thymidine uptake.

The results of the study are summarized in Figs. 8 and 9. As shown, tuftsin highly potentiates the immunogenic activity of antigen-fed macrophages with maximum influence at concentrations of about $5\,\mu$M. Intracellular protein and RNA syntheses were also augmented during cell activation (70). The stimulation by tuftsin was rather specific and markedly effected by alterations in the peptide's structure. Similar to findings in other assay systems, the synthetic analogs tested were divided into three groups: active derivatives some of which rather potent; [Ala^1]tuftsin and the pentapeptide tuftsinylglycine, even more than tuftsin; inactive compounds and inhibitors (Figs. 8 and 9). Comparison, however, of results of immunostimulation with those of phagocytosis or NBT-reduction experiments reveal three major discrepancies. First, the dose-response patterns of immunostimulation usually reach a maximum and is followed by a sharp decrease of activity at higher concentrations, whereas response curves obtained from phagocytosis experiments have a typical saturation characteristic (Figs. 8 and 9). Second, several contradictions exist between activities of certain analogs in the two assay systems. This is particularly

evident with [Ala^1]tuftsin, whose immunostimulatory capacity exceeds that of tuftsin, but exhibits a powerful inhibitory effect on phagocytosis or NBT-reduction. Close examination of the various cell stimulations by tuftsin analogs indicate that the integrity of a minimal sequence, Pro-Arg, is essential for activity. The dipeptide L-prolyl-L-arginine indeed considerably stimulated immunogenic activity of macrophages, though with a different dose response dependence than tuftsin (Fig. 8). It did not, however, show any effect upon phagocytosis or NBT-reduction (63). It is possible, that two different receptors are involved that share some common features (e.g., recognition of Pro-Arg) but are specifically functionalized by certain additional sequence. We do not have, as yet, any indications which can substantiate this latter assumption. The third discrepancy concerns the concentration of tuftsin needed to elicit maximal immunopotentiating effects which is about 10 times higher than the corresponding amounts required to obtain maximal phagocytosis. Again, these difference may stem from interactions with two different receptors. It may also be attributed to the prolonged incubation period of macrophages with tuftsin in the immunostimulation assay, during which the peptide may be considerably degraded by proline-peptidases (20).

Aiming to explore the possibility that tuftsin may play an *in vivo* immunostimulatory role, Florentin *et al.* have examined its effect on both humoral and cell-mediated immune responses (71). Injecting mice with tuftsin, at different time intervals before immunization, caused an augmentation of antibody responses both to thymus-dependent (TNP-KLH) and to thymus-independent (TNP-LPS) antigens. The time dependence of the two events is, however, different. Thus, reaction to the latter antigen was stimulated rapidly (1 day) and lasted for a short time (3 days) subsequent to tuftsin administration. Potentiation of the antibody response to the TNP-KLH antigen, on the other hand, occurred only 3 days after tuftsin injection. Florentin et al. have suggested that this may result either from stimulation of T helper cell function, or from antigen-processing by macrophages in addition to the B cell stimulation (71).

Two other important types of immune responses potentiated by tuftsin were demonstrated by Florentin *et al.* (71); a) activation of the cytostatic effect

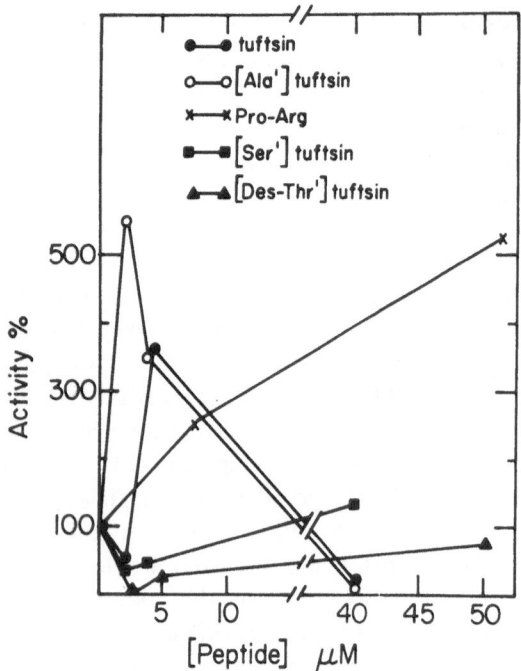

Fig. 8. Effects of tuftsin, its NH$_2$-terminal analogs and Pro-Arg on the immunogenic activity of antigen pulsed macrophages (68–70).

exerted by macrophages on tumor cells (L1210 leukemic cells) and b) an enhancement of antibody-dependent cell cytotoxic (ADCC) activity of spleen cells. These immune mechanisms were already

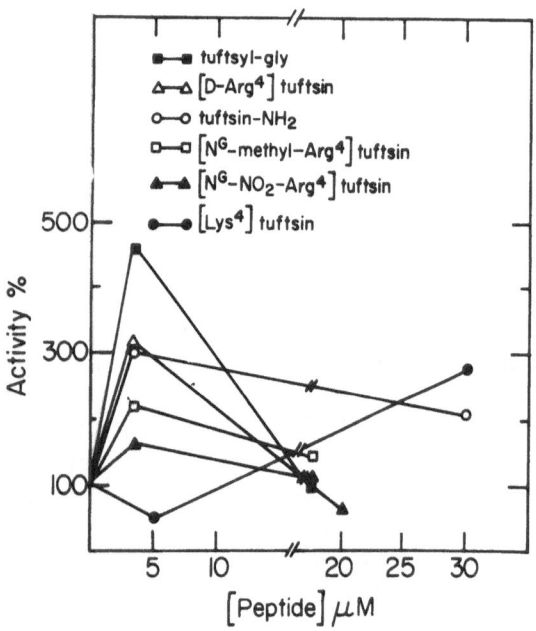

Fig. 9. Effects of COOH-terminal analogs of tuftsin on the immunogenic activity of antigen pulsed macrophages (68–70).

proved to be of importance in the control of tumor growth, suggesting therefore the possible application of tuftsin in cancer immunotherapy.

Anti-tumor effect of tuftsin

In direct line with the above study, the work of Nishioka (72) indicates that tuftsin may probably act as an immunotherapeutic agent. Thus, mice inoculated, intraperitoneally, simultaneously with leukemic cells (L1210; 10^4 cells) and tuftsin (0.2 μg) had a statistically significant mean survival time, which is longer than the control group.

In the same study it was also demonstrated that tuftsin markedly enhanced *in vivo* spreading of macrophages as well as augmentation, of *in vitro* cytotoxicity of macrophages though at a rather moderate level (72).

The *in vivo* anti-tumor effect of tuftsin on yet another type of tumor was recently demonstrated by Nishioka and coworkers (73). Thus, melanoma cells (Cloudman S-91) were injected i.p. into mice leading to significant tumor colonies in the lung 4 weeks post inoculation. Single treatment with tuftsin, tested over a wide range of concentrations, either i.p. or i.v. affected tumor growth only slightly. On the other hand, multi-administrations of tuftsin, 3 times weekly, starting 1 day after tumor inoculation led to a marked decrease (5-fold) in the number of melanoma colonies. The effect of tuftsin was prominent and equally effective in doses of 10 to 10^3 μg per mouse. Tuftsin therapy that started as late as 14 days after tumor injection was still significantly effective. It is most noticeable that no side effects were detectable due to tuftsin's treatment at doses of 0.1–10^4 μg/animal every other day up to six weeks (73).

Increase of cell migration by tuftsin

Enhancement of cell migration by tuftsin was assessed on PMNL and monocytic cells by various methods. Hoshmanheimo *et al.* (74), using a modified version of leukocyte migration inhibition test (LMAT), have demonstrated that tuftsin significantly increases random migration in a mononuclear cell preparation of lymphocytes and monocytes. Since lymphocytes are devoid of response to tuftsin, one can attribute this activity to the mononuclear fraction of cells alone. Similar findings in which the effect of tuftsin is tested on capillary tube migration inhibition test, were also reported (75). In the latter case both normal as well as sensitized human mononuclear cells displayed the same effect.

Results obtained using PMNL cells are less clear and are contradictory. Nishioka et al. (11) tested the effect of tuftsin, using capillary tubes, and clearly demonstrated that random migration is increased in PMNL cells. Using the same type of cells and the LMAT test, no effect could be found by Horsmanheimo and collaborators (74). Goetzl and coworkers have even reported that tuftsin has an inhibitory effect on random migration of these cells (76). We cannot explain these discrepancies, although Najjar (13) has suggested that these differences may stem from impurities, found at times, in tuftsin preparations. It should be noted that both monocytes and PMNL cells have tuftsin receptor sites and both are stimulated to phagocytose by tuftsin. It therefore seems likely that tuftsin does induce random migration in PMNL as it does in monocytes.

Nishioka (75) has studied the effect of tuftsin on sensitized cells obtained from patients with malignant melanoma. The effect of tuftsin on migration inhibition induced by melanoma antigen, was to increase migration of mononuclear cells. The ability of tuftsin to override migration inhibition antigen material, obtained from another source, was also demonstrated. The author has suggested that aberrations often observed in migration inhibition tests may stem from unwarranted release of tuftsin from its carrier molecule.

In a recent abstract, tuftsin was reported to enhance migration and abolish inhibitory activity by migration inhibition factor (77). These findings are in accordance with those of Nishioka.

In vivo Anti-bacterial activity of tuftsin

As indicated in previous sections, the antibacterial capacity of tuftsin has been quite extensively assessed in the in vitro systems. Only one detailed study, by Martinez et al., discuss various parameters related to the in vivo effects of tuftsin in infected animals (30).

We have described, the stimulatory effect of tuftsin on the respiratory-burst, i.e. bactericidal machinary, in PMN-leukocytes. Martinez and co-

workers have studied the bactericidal activity of peritoneal macrophages of mice injected intraperitoneally with Listeria monocytogenes with or without tuftsin. Five minutes following injection the animals were sacrificed, peritoneal macrophages harvested and then subjected to appropriate incubation at 37 °C. At various time intervals intracellular viable bacteria was determined. As shown in Table 3 the effect of tuftsin was very significant. Controls revealed after 15 min of incubation, only 5% killing of bacteria, whereas in the presence of tuftsin, 10 or 20 mg per kg body weight, as high as 50% and 69% of bacteria were killed, respectively. After 30 min, bacterial killing amounted with tuftsin to respective numbers of 63% and 72% as compared with 20% of control. Tuftsin exhibited similar effect, though somewhat smaller, on bacterial killing in infected leukemic mice (30).

As an additional parameter of anti-bacterial activity, Martinez et al. have studied the blood clearing effect of tuftsin in infected mice (30). Normal mice were injected intravenously with different bacteria, Listeria monocytogenes, E. Coli and Serratia marcescens, followed by immediate intraperitoneal injection of tuftsin at doses of 10 or 20 mg per kilogram weight. At various time intervals, blood was taken through cardiac puncture and its bacterial content evaluated. As shown in Fig. 10 for clearance of Staph. Aureus, and of E. Coli, the initial effect of tuftsin is dramatic. After 60 min, however, counts of bacteria in control are close to those in the presence of tuftsin. Similar clearance patterns were described for other bacteria studied. Increased blood clearing of bacteria in leukemic mice was also observed in the presence of tuftsin, it was, however, lower when compared to the corresponding effect in normal mice.

Finally, Martinez has shown that animal survival when infected with pathogenic microorganism can be increased by tuftsin (78). Thus, 20 mg/kilo of tuftsin when injected i.p. to mice treated with nearly lethal dose of pneumococcus increased animal survival from 10% in control to 50%.

Tuftsin's related sequences included in peptides and proteins: coincidence or functional significance?

Peptides

Sequences related to tuftsin in which a proline

Table 3. The effect of tuftsin on the intracellular bactericidal activity of mouse peritoneal macrophage using Listeria monocytogenes (Martinez *et al.* (30)).

Assay lot	Time (minutes)	$K_t{}^a$	% Bacteriacidal activity[b]
	0	–	–
Control	15	0.022	5.2
	30	0.096	20.0
	60	0.152	29.6
Tuftsin	0	–	–
10 mg/kg	15	0.29	49.9
i.p.	30	0.431	63.0
	60	0.440	63.78
Tuftsin	0	–	–
20 mg/kg	15	0.503	68.6
i.p.	30	0.552	71.3
	60	0.567	72.1

[a] $K_t = \log N_0/N_t$, where N_0 is the number of intracellular bacteria at t_0 and N_t the number of intracellular bacteria at time t.

[b] Calculated as $\dfrac{N_0-N_t}{N_0}$, where N_0 = number of bacteria at time 0, and N_t = number of bacteria at time t.

residue is situated in close proximity to either or both basic amino acids, lysine and arginine, are a frequent occurrence in biologically active peptides. The peptides vasopresin, neurotensin, substance P, bradykinin and adrenocorticotropin hormone, all contain such moieties. The chance occurrence of

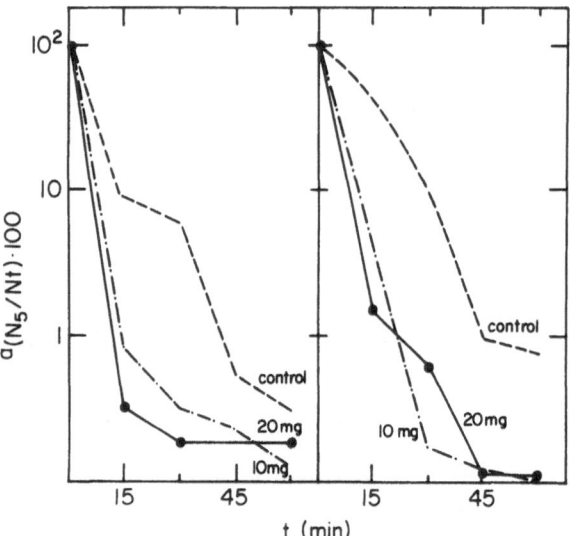

Fig. 10. The effect of tuftsin on bacterial clearing: (Left) Staph. aureus; (Right) E. coli (Martinez *et al.* (30)).

such structural similarities in these diversely active peptides is beyond statistical probability. Often, especially in high doses, functional cross-reactivity exist between different peptides (79–83). Tzehoval *et al.* (68) have shown, for example, that the 'anti-allergic peptide', Asp-Ser-Asp-Pro-Arg, luteinizing hormone-releasing hormone (LH-RH) and bradykinin have immunostimulatory effects.

The undecapeptide Substance P (SP) has a widespread distribution in the vertebrate nervous system. It possesses hypotensive, vasodilatatory and smooth muscle contracting properties and perhaps plays a role in sensory neurotransmission (84–86). The N-terminal tetrapeptide of SP is H-Arg-Pro-Lys-Pro- is a sequence which bears certain similarities to tuftsin. This fact inspired Bar-Shavit *et al.* (55) to explore the possibility that SP and its N-terminal tetrapeptide may exhibit tuftsin-like activity. Indeed, these peptides displayed strong phagocytosis stimulating activity (Fig. 11). Moreover, both peptides can specifically bind to tuftsin's receptor sites on mouse normal and thioglycollate-stimulated macrophages and human PMN-leukocytes (Fig. 12). It was therefore suggested that the phagocytosis enhancing activity of Substance P may be of relevance in inflammatory processes of neural origin (55).

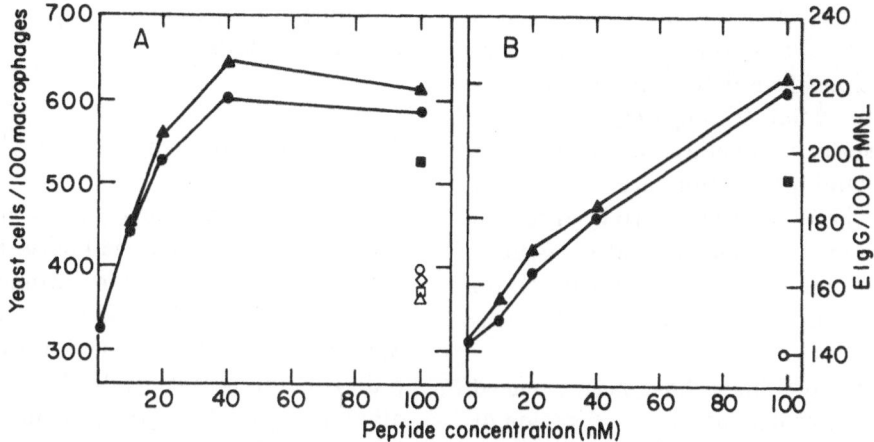

Fig. 11. The effect of substance P (SP) and related peptides on the phagocytosis of: A – Yeast cells by normal macrophages, and B – Sheep erythrocyte-IgG by human PMNL. ●, SP; ▲, N-terminal tetrapeptide; ■, tuftsin; o, C-terminal heptapeptide; △, C-terminal octapeptide; □, Arg-Pro; ◊, Lys-Pro (55).

The above findings have led us once again to the possibility that 'general-receptors' for Arg-Pro and related sequences exist. In addition, one must seriously question, the multitude of *in vitro* functions to which tuftsin or its 'sequence-relatives' has been implicated. To what extent, do these peptides, participate in the various endocytic processes *in vivo* and are their activities only confined to professional phagocytic cells will undoubtedly be the subject of future research.

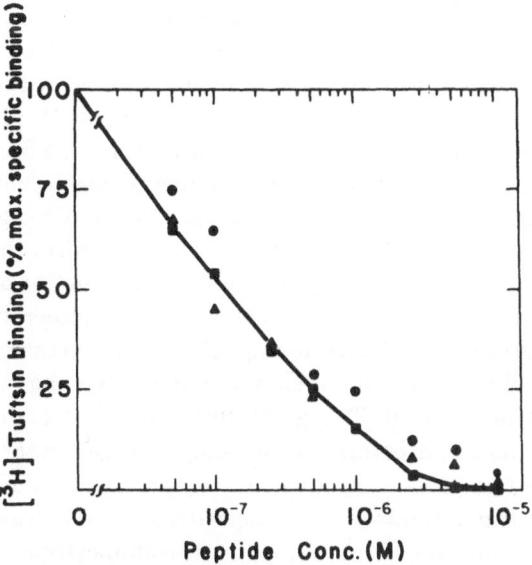

Fig. 12. Effect of substance P (SP) and related peptides on the binding of [³H]-tuftsin to thioglycollate-stimulated macrophages. ●, SP; ▲, N-terminal tetrapeptide; ■, unlabeled tuftsin. The line represents the competition exhibited by unlabeled tuftsin (55).

Proteins

Sequences related to tuftsin within proteins are quite rare. The few known examples include the H5 histone, -Lys-Pro-Lys at location 10–12 from the amino terminal (87); histone 4, contains the sequence -Thr-Lys-Pro (88); C-reactive protein, contain -Lys-Pro-Arg and twice the -Thr-Lys-Pro sequence (89); the phosphoprotein P12 from Raucher murine leukemia virus (R-MuLV) which contains the whole tuftsin sequence -Thr-Lys-Pro-Arg- at positions 9–12 from the protein's amino terminal (87, 90).

The presence of a tuftsin-moiety within the P12 protein and the effect of tuftsin on the degradation of the viral *gag* gene products P65-70 by proteolytic factor, motivated Luftig *et al.* to study the effect of tuftsin on the MuLV infected cell (91) (also see this issue). Tuftsin added in concentrations up to 100 µM to cultures of these cells resulted in a three fold increase in virion associated reverse transcriptase. It was suggested that the P12 protein may be a stimulator of cell's membrane activity thus leading to increased virion production. It is perhaps of relevance to mention here that the study of Martinez (78) which showed that the survival of mice inoculated with myxovirus was only 10% in presence of tuftsin as compared to 40% of control. The inceased lethality in animals may be due to increased phagocytosis, but not killing, of virus. It may also stem, however, from augmentation of viral production by tuftsin! Similar induction of endogenous murine retrovirus by tuftsin or by a related analog Thr-Pro-Arg-Lys (kenstin), at peptides concentrations from 0.01 to 100 μg/ml, was recently reported by Suk & Long (92).

It is still questionable whether the above effects of tuftsin on viral production are membrane-dependent and whether they have certain structural specificities or not. It is not illogical to assume, however, that these and other activities of tuftsin do have some common, as yet undeciphered, mechanism of action.

Certain remote homologies have been detected between C-reactive protein and the C_H2 region of IgG i.e. the origin of tuftsin (89). It is also possible that viral P12 evolved from H5 histones or related protein through a genetic recombination involving only the amino terminal end of the H5 histones (87). No correlation, however, between the activity of these proteins and that of tuftsin has so far been demonstrated.

Clinical aspects of tuftsin

The increased susceptibility of humans, particularly children, to severe and fulminant infections after splenectomy is well documented e.g. (93–96). This susceptibility is primarily evident after elective splenectomy but is also common after traumatic operations (97). The reason for the heightened rate and severity of infection in asplenic individuals is not clear (98). Earlier, we described the studies of Najjar and coworkers in which leukokinin obtained from splectomized patients lacks both the capacity to stimulate phagocytosis of phagocytic cells and the ability to yield free tuftsin upon digestion with leukokininase or trypsin. Based on these considerations, the spleen was assigned to actively participate in leukokinin processing to yield tuftsin. This also led to the discovery and characterization of a familial tuftsin deficiency (14, 99, 99a). It has been established that people who suffer from either acquired or familiar tuftsin deficiency display a direct correlation between their sensitivity toward bacterial infections and the quality of their leukokinin i.e. tuftsin content (14, 99a).

Recently, an antitumor effect of tuftsin has been described (72, 73). Though only preliminary results are available it should not be entirely and immediately excluded that combatting tumors *via* the

reticuloendothelial system is part of the physiological role of tuftsin.

Radioimmunoassay for tuftsin and its applications

In view of the above, it is obvious that an accurate and sensitive method of evaluating tuftsin's level in human blood would be of utmost clinical-diagnostical significance. Spirer *et al.* (17) have rendered tuftsin antigenic through coupling of p-diazonium phenylacetyl-tuftsin to bovine serum albumin (BSA). Using the BSA-tuftsin conjugate to immunize rabbits antisera were obtained that bound ^{125}I-labeled p-amino-phenylacetyl-tuftsin at dilutions up to 1:1500. Binding of the radiolabeled peptide was inhibited by tuftsin and some of its synthetic analogs, but was not affected by various unrelated peptides (17). Using the radioimmunoassay, quantitative determination of material immunochemically related to tuftsin was performed in the sera of normal and various splenectomized subjects, after short treatment of the sera with trypsin (17, 100). The results are depicted in Fig. 13. The significant difference between the average levels of tuftsin found in the blood of normal humans (35 subjects), 255.7 ± 10.2 ng/ml, and the amount detected in patients who underwent elective splenectomy, 118.0 ± 7.9 ng/ml (20 subjects operated for a variety of pathological reasons) and 126.05 ± 10.2 ng/ml (38 subjects operated due to Hodgkin's disease) was anticipated as indicated in the phagocytosis experiments of Najjar (14, 21, 99, 99a). The rather normal tuftsin's levels, 234.3 ± 10.2 ng/ml, found in the traumatic group (50 subjects) were also accounted for by Najjar's suggestion (21) that splenic tissue is implanted in the abdomen during the spleen rupture and function as mini-spleens. Two additional cases (not shown in figure) are worth mentioning: a 7-month-old infant in whom a spleen torsion was diagnosed had a tuftsin level of 98 ng/ml three days after the incident and before splenectomy, but two weeks and two months after splenectomy showed values of 50 ng/ml and 40 ng/ml respectively. A 5-year-old girl suffering from idiopatic thrombocytopenic purpura and recurrent pyogenic skin and bronchopulmonary infections, without previous evidence of impaired cellular or humoral immunity, showed value of 107 ng/ml of tuftsin in her blood.

A group of six patients was studied immediately after Hodgkin's disease was diagnosed and before splenectomy was performed, showing a mean tuftsin level of 246.2 ± 8.3 ng/ml and thus not significantly different from normal levels. The finding are consistent with the fact that in Hodgkin's disease splenic function remains intact despite the attack on the reticuloendothelial system (101). They suggest that the practice of performing staging laparotomy and splenectomy in patients with the disease should be reconsidered (100). The radioimmunoassay studies indicate that tuftsin is probably not free in circulation but rather bound to leukokinin. The determination of potential tuftsin's levels in human serum, proved to be reproducible and with a high correlation index (r = 0.98). The assays may perhaps be important in managing patients in hyposplenic states. The results may help to determine the necessity and optimum duration of prophylactic antibiotic treatment.

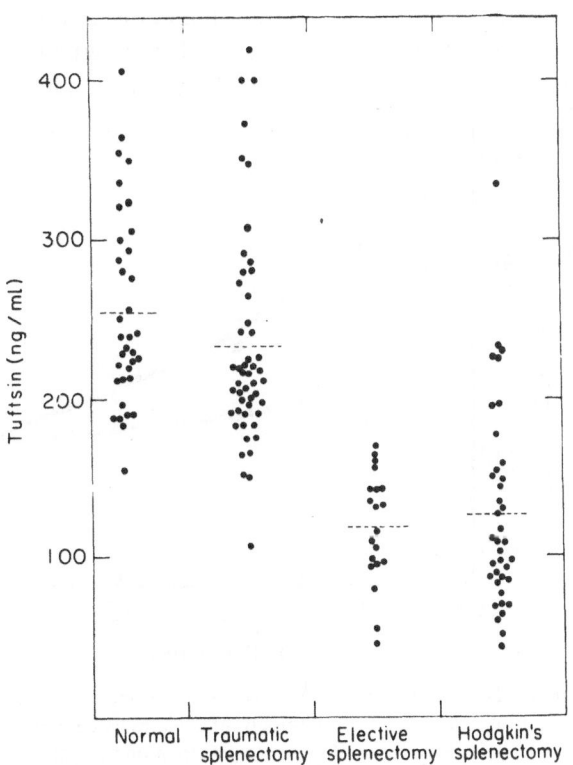

Fig. 13. Amounts of tuftsin, as determined by radioimmuno-assay, in trypsinized serum of healthy normal and splenectomized patients. Dotted lines indicate mean values for each set of experiments (17, 100).

Tuftsin has no apparent toxicity in vivo

In view of the data presented above it seems that the use of tuftsin as a general anti-infectious drug should be considered. As a prerequisite in evaluating the use of tuftsin as a therapeutic agent, its possible toxicological effects must be examined. Thus, Najjar has injected tuftsin intravenously to rats up to the amounts of 25 mg per kilogram of body weight. Monitoring of blood pressure, heart rate, respiratory rate and electrocardiograms, no change in these parameters was noted upon tuftsin administration (13). Preliminary results from our laboratory (unpublished) indicate that the lethal-dose-50 of tuftsin when injected intravenously to mice is about 2.5 gram per kilo body weight. Upon intraperitoneal injection of tuftsin up to 2.5 gram per kilo body weight mice survival was 100%.

On the mode of tuftsin's action

The existance of specific receptor sites for tuftsin on surfaces of phagocytic cells and the direct correlation between binding parameters and the amounts of peptide required to elicit high cellular responses has already been discussed above. It is logical to assume that the action of tuftsin is indeed receptor-mediated. Constantopoulos and Najjar have shown that the presence of sialic acid on the surface of PMN-leukocytes is essential for tuftsin's stimulatory effect. Treatment of cells with bacterial neuraminidase, which leave the cells viable, lead to the abolishment of both tuftsin's and leukokinin's effects (53).

It is very likely that the first encounter between the cell and tuftsin result from the attraction of cell's negative charge created by its multimillion (135×10^6) sialic acid moieties (53) and the positive charge of the basic tetrapeptide. Following or during attachment tuftsin may assume a certain conformation and fits itself into a specific locus of action. How the membrane perturbation by tuftsin is translated into cellular function remains unclear. The possibility that tuftsin transfers its signal *via* second messenger was investigated. As an obvious initial approach, Stabinsky et al. have studied the effect of tuftsin on intracellular levels of cAMP and cGMP in phagocytes (102). It was found that incubation of tuftsin (2.5×10^{-7} M) with either

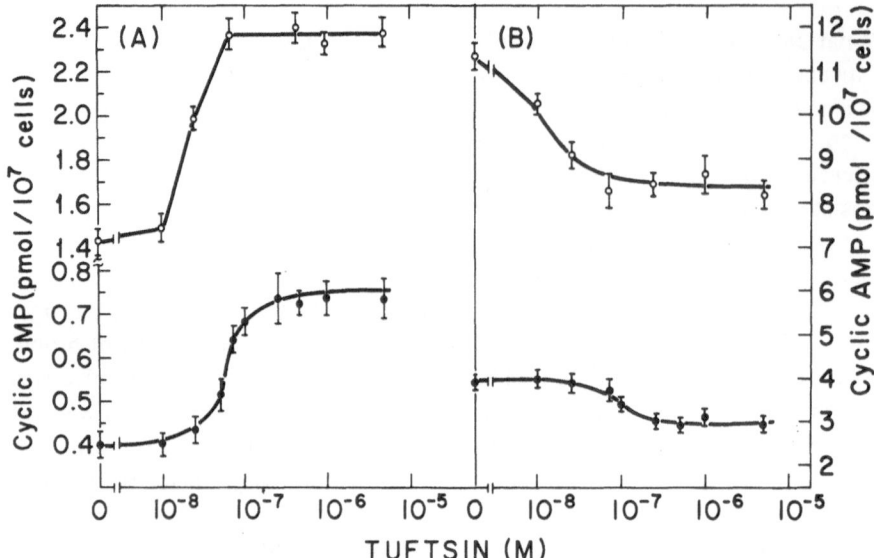

Fig. 14. The dependence of intracellular levels of cGMP (A) and cAMP (B) in normal human PMNL (●) and thioglycollate-stimulated mouse peritoneal macrophages (○) on tuftsin concentration (102).

human PMN leukocytes or thioglycollate-stimulated mouse peritoneal macrophages resulted in the increase of 80–90% in intracellular cGMP levels, accompanied by a decrease of 20–25% in intracellular cAMP levels. The effects were detectable after 4 min of incubation, reached a maximum at 10–20 min and persisted for at least 60 min. The dose-response patterns are illustrated in Fig. 14. The concentration dependences of the stimulatory effects of tuftsin on modulation of intracellular levels of cyclic nucleotides and on phagocytosis are similar, suggesting a cause and effect relationship between the two phenomena (102). This mutual but opposite involvement of the two cyclic nucleotides in the phagocytosis process is further supported by the finding that exogenous addition of the cell-penetrable 8-Br-cGMP and 8-Br-cAMP exerted stimulatory and inhibitory effect, respectively, on the process (102). Moreover, it was shown that other related cellular responses such as locomotion (103, 104), chemotactic responsiveness (103, 104), NBT-reduction (105) and granule discharge (106, 107) are inhibited in granulocytes by exogenous agents that increase intracellular levels of cAMP, whereas agents that increase cGMP stimulate these processes (104–110).

From available literature, it seems likely that increases in intracellular cGMP levels result, at least in some cases, from the influence of membrane active agents on the activity of cellular guanylate cyclase (111). It has been suggested that one possible mechanism by which a biologically active compounds promote increases of cGMP levels involves changes in the intracellular concentration or distribution of Ca^{++} (112). Calcium activation of guanylate cyclase and cGMP accumulation,

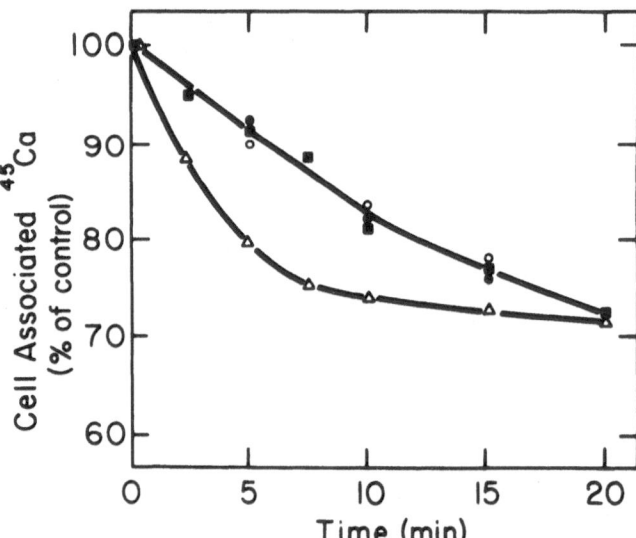

Fig. 15. Time dependence of the effect of tuftsin and its analogs on outflux of $^{45}Ca^{2+}$ from thioglycollate-stimulated mouse peritoneal macrophages. Macrophages (in monolayers) pre-loaded with $^{45}Ca^{2+}$ were incubated in the absence (■-control) or presence of the tested peptide (2.5×10^{-7} M) for the indicated time intervals. Each point is the mean of quadruplicate determinations.

(△-tuftsin; ○-[N-Acetyl-Thr¹]tuftsin; ●-[Des-Thr¹]tuftsin) (102).

have subsequently been demonstrated, for a number of agents in several tissues, e.g. (113, 114). Stabinsky et al. have found that $^{45}Ca^{++}$ influx into PMN-leukocytes and macrophages is not effected by tuftsin (102). The peptide did, however, enhance the release of $^{45}Ca^{++}$ from preloaded cells in a rather specific manner (Fig. 15). Therefore one can not exclude the possibility that the initial effect of tuftsin is to release exchangeable cellular Ca^{++}. The role of Ca^{++} in regulating cellular functions is not entirely clear, but it is possible that the cation may stimulate, directly or indirectly, both guanylate cyclase and cAMP phosphodiesterase activities (111). Alternatively, cGMP, may by itself, selectively stimulate the activity of cyclic nucleotide phosphodiesterase that catalyzes cAMP hydrolysis (115).

The above mentioned results are rather preliminary and should be looked at more closely and perhaps also in relation to the effect of tuftsin on cellular contractile elements. It seems plausible, however, that tuftsin's augments phagocytes activity *via* its effect on intracellular cyclic nucleotide levels, be it mediated through a primary effect on Ca^{++} associated with cellular components or *via* another, as yet, unknown mechanism.

Some concluding remarks

We have tried to outline in a panoramic and rather concise manner the chemical and biological characteristics of tuftsin. It was not intended to be an exhaustive review however, most of the published relevant literature are included. According to the authors view-point, several topics have been more deeply covered, whereas others are only briefly mentioned or referred to.

Tuftsin possesses a wide and diverse spectra of biological activities which is exerted primarily on or *via* professional phagocytic cells. The various effects of tuftsin are directly dependent upon strict architectural requirements and are initiated, at least in part, by provoking specific cellular receptors.

Tuftsin is a small peptide, Thr-Lys-Pro-Arg. Although three of its constituent amino acids, Thr, Lys and Arg, are tri-functional and the forth, Pro, has unique conformational implications, it is not logical that the amount of 'information' stored in this sequence can account for all its functions. Is

tuftsin a mere general effector agent capable of perturbing certain membranes? Is it possible that all of the cellular processes stimulated by tuftsin are triggered by the same signal? What is, if any, the true physiological role(s) of tuftsin *in vivo*?

Future work on tuftsin will be undoubtedly aimed at solving these questions, as well as, at understanding the various other parameters concerning, its biological functions and potential clinical application. The main research goals will be an extension of current studies and probably include: further elucidation of the mode of action of tuftsin on the cellular and molecular levels; attempted isolation and characterization of specific receptors from phagocytic cells; studies on the processing of leukokinin to yield tuftsin and its possible relation to pathological situations of recurrent infectious diseases; elaboration of the *in vivo* immunostimulatory effect of tuftsin on both humoral and cell-mediated immune response; examination of the possibility that tuftsin may play *in vivo* roles as a recruiter of phagocytic cells and as a chemotactic agent; extended radioimmunoassay studies to further establish the direct correlation between tuftsin's levels and impaired splenic functions; *in vivo* broad-range anti-infectious and antitumor studies including humans and last but not least, synthetic efforts directed toward producing highly-active and long-acting analogs.

In the introduction we have referred to the findings of Najjar and co-worker that leukokinin, the precursor protein of tuftsin, is only one member in a group of cytophilic immunoglobulins with well-defined affinities to various blood cells. The possibility that tuftsin is the first representative of a family of peptides which are derived from parental immunoglobulins which act specifically on erythrocytes, thrombocytes or other cells, should be considered.

Addendum

Recent studies on the antitumor activity of tuftsin performed by Catane *et al.* (116) are described below. Tuftsin was found to be active against 3-methyl-chloranthrene induced transplantable fibrosarcoma in C3H mice. When 10^5 tumor cells were injected intraperitoneally (IP) to control mice there was a 100% take and tumor was lethal

94

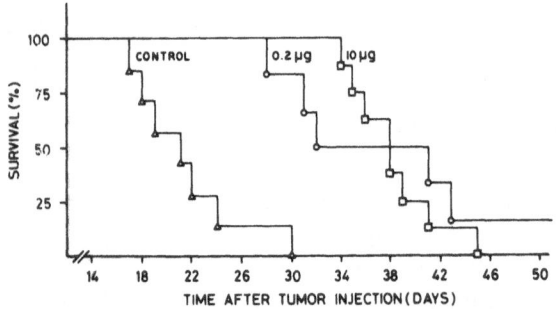

Fig. 16. Antitumoral activity of tuftsin on IP induced fibro-sarcoma.

with a median survival (MS) of 21 days. Marked prolongation of survival was observed when tuftsin was introduced IP, 3 times per week, starting on the day of tumor inoculation (Fig. 16). A dose range of 0.2–500 μg/kg gave a MS time of 39 days and 20 percent of the mice did not develop tumors for a period of 80 days or more.

When tumor cells were injected subcutaneously, the effect of tuftsin administered IP was measured in terms of tumor size. Although the effect is less marked, Fig. 17 demonstrates that tuftsin retards tumor growth. At a concentration of 500 μg/kg the most dramatic effect was observed on day 30, where as approximate four fold difference in tumor volume is observed in comparison to control. It should be noted that this sharp difference in tumor size is practically lost with time.

Currently, various analogs are being tested in this system. [Ala¹] tuftsin, an inhibitor of tuftsin's phagocytic activity, neither inhibits nor enhances anti-tumor activity measured both in terms of survival and tumor size.

The anti-tumor activity of tuftsin appears to reside in macrophages. Not only do receptor sites for tuftsin exist on macrophages (see this review), but carrageenan, an inhibitor of macrophage activities, completely suppresses anti-tumoral effects of tuftsin.

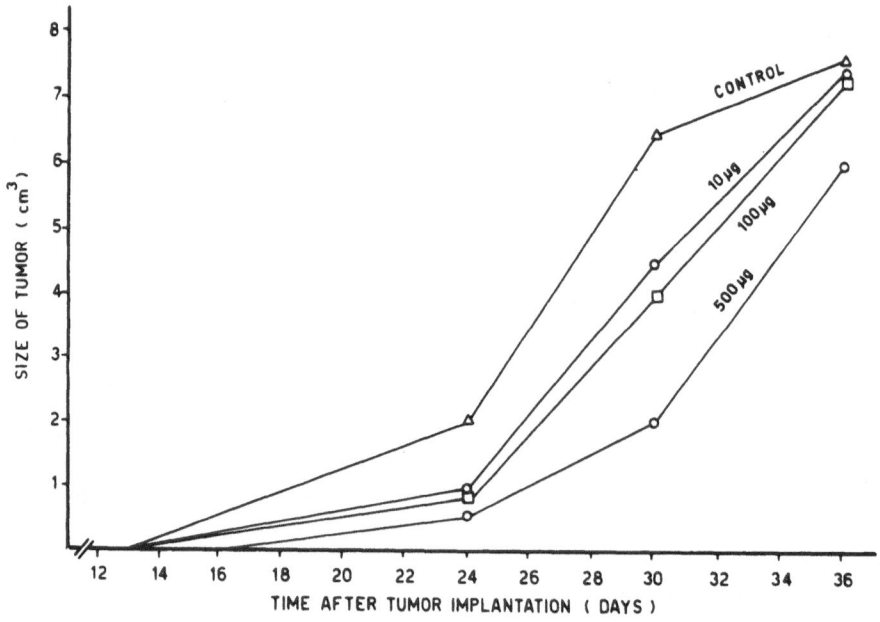

Fig. 17. The effect of various tuftsin concentrations on subcutaneously induced tumors.

References

1. Fidalgo, B. V. & Najjar, V. A., 1967. Proc. Natl. Acad. Sci. USA 57: 957–964.
2. Fidalgo, B. V., Katayama, Y. & Najjar, V. A., 1967. Biochemistry 6: 3378–3385.
3. Fidalgo, B. V. & Najjar, V. A., 1967. Biochemistry 6: 3386–3392.
4. Thomaidis, T. S., Fidalgo, B. V., Harshman, S. & Najjar, V. A., 1967. Biochemistry 6: 3369–3377.
5. Najjar, V. A., Fidalgo, B. V. & Stitt, E., 1968. Biochemistry 7: 2376–2379.
6. Najjar, V. A., 1974. Advances in Enzymology (Meister, A., ed.), Vol. 41, pp. 129–178, John Wiley and Sons Inc., New York.
7. Constantopoulos, A. & Najjar, V. A., 1974. Eur. J. Biochem. 41: 135–138.
8. Najjar, V. A. & Nishioka, K., 1970. Nature 228: 672–673.
9. Nishioka, K., Constantopoulos, A., Satoh, P. S. & Najjar, V. A., 1972. Biochem. Biophys. Res. Commun. 47: 172–179.

10. Nishioka, K., Constantopoulos, A., Satoh, P. S., Mitchell, W.M. & Najjar, V. A., 1973. Biochim. Biophys. Acta 310: 217–229.
11. Nishioka, K., Satoh, P. S., Constantopoulos, A. & Najjar, V. A., 1973. Biochim. Biophys. Acta 310: 230–237.
12. Najjar, V. A., 1980. Advances in Experimental Medicine and Biology (Escobar, M. and Freidman, H. eds.), 121A: pp. 131–148, Plenum Press, New York.
13. Najjar, V. A., 1979. The Reticuloendothelial System: A Comprehensive Treatise (Sbarra, A. J. and Strauss, R. R., eds.), Vol. II, Biochemistry of the Reticuloendothelial Sytem, Plenum Press, New York, in press.
14. Najjar, V. A., 1978. Exptl. Cell Biol. 46: 114–126.
15. Satoh, P. S., Constantopoulos, A., Nishioka, K. & Najjar, V. A., 1972. Chemistry and Biology of Peptides (Meinhofer, J., ed.), pp. 403–408, Ann Arbor Science Publisher, Michigan.
16. Edelman, G. M., Cunningham, B. A., Gall, W. E., Gottlieb, P. D., Rutishauser, U. & Waxdal, M.J., 1969. Proc. Natl. Acad. Sci. USA 63: 78–85.
17. Spirer, Z., Zakuth, V., Bogair, N. & Fridkin, M., 1977. Eur. J. Immunol. 7: 69–74.
18. Sequences of Immunoglobulin Chains (Kabat, E. A., Wu, T. T. & Bilofsky, H., eds.), U.S. Department of Health, Education and Welfare. National Institutes of Health Publication No. 80–2008.
19. Skeggs, L. T., Jr., Dorer, E. F., Kahn, J. R., Lentz, K. E. & Levine, M., 1976. Amer, J. Med. 60: 737–748.
20. Rauner, R. A., Schmidt, J. J. & Najjar, V. A., 1976. Mol. Cell. Biochem. 10: 77–80.
21. Najjar, V. A. & Constantopoulos, A., 1972. J. Reticuloendothel. Soc. 12: 197–215.
22. Stabinsky, Y., Fridkin, M., Zakuth, V. & Spirer, Z., 1978. Int. J. Peptide Protein Res. 12: 130–138.
23. Yajima, H., Ogawa, H., Watanabe, H., Fujii, N., Kurobe, M. & Miyamoto, S., 1975. Chem. Pharm. Bull. 23: 371–374.
24. Mishimura, O. & Fujino, M., 1976. Chem. Pharm. Bull. 24: 1568–1575.
25. Fridkin, M., Stabinsky, Y., Zakuth, V. & Spirer, Z., 1976. Peptide 1976, Proc. 14th Eur. Peptide Symp. (Loffet, A., ed.), pp. 541–550, Press Universitaire de Bruxelles, Bruxelles.
26. Vicar, J., Gut, V., Fric, I. & Blaha, K., 1976. Collec. Czech, Commun. Chem. 41: 3467–3473.
27. Fridkin, M., Stabinsky, Y., Zakuth, V. & Spirer, Z., 1977. Biochim. Biophys. Acta 496: 203–211.
28. Nozaki, S., Hisatsune, K. & Muramatsu, I., 1977. Bull. Chem. Soc. Japan 50: 422–424.
29. Konopinska, D., Nawrocka, E., Siemion, I. Z., Slopek, S., Szymaniec, St. & Klonowska E., 1977. Int. J. Peptide Protein Res. 9: 71–77.
30. Martinez, J., Winternitz, F. & Vindel, J., 1977. Eur. J. Med. Chem.-Chim. Ther. 12: 511–516.
31. Stabinsky, Y., Fridkin, M., Zakuth, V. & Spirer, Z., 1977. Israel Chem. Soc. Proceedings 44th Annual Meeting, Abst. MN-10.
32. Chaudhuri, M. K. & Najjar, V. A., 1979. Anal. Biochem. 95: 305–310.
33. Stabinsky, Y., Gottlieb, P. & Fridkin, M., 1980. Mol. Cell. Biochem., 30: 165–170.
34. Sucharda-Sobczyk, A., Siemion, I.Z. & Konopinska, D., 1979. Eur. J. Biochem, 96: 131–139.
35. Konopinska, D., Siemion, I.Z., Szymaniec, S. & Slopek, S. 1979. Polish. J. Chem. 53: 343–351.
36. Stern, M., Warshawsky, A. & Fridkin, M., 1979. Int. J. Peptide Protein Res. 13: 315–319.
37. Yasumura, K., Okamoto, K. & Shimamura, S., 1977. Yakugaku Zasshi, 97: 324–329.
38. Nozaki, S., Kobayashi, K., Hisatsune, K. & Muramatsu, I., 1976. Peptide Chemistry, Proc. 14th Japanese Symp. (Nakajima, T., ed.), pp. 131–134, Protein Research Foundation, Osaka.
39. Hisatsune, K., Kobayashi, K., Nozaki, S. & Muramatsu, I., 1978. Chem. Pharm. Bull. 26: 1006–1007.
40. Konopinska, D., Nawrocka, E., Siemion, I. Z., Szymaniec, S. & Slopek, S., 1978. Arch. Immunol. Ther. Exper. 26: 151–157.
41. Matsuura, S., Takasaki, A., Hirotami, H., Kotera, T. & Fujiwara, S., 1975. Japan Kokai 13: 373; Chem. Abstr. 1975, 83, 114937.
42. Okamoto, K. & Simamura, S., 1976. Yakugaku Zasshi. 96: 315–320.
42a.Konopinska, D., Nawrocka, E., Siemion, I. Z., Szymaniec, S. & Slopek, S., 1976. Peptides 1976 (Loffet, A., ed.), pp. 535–539, Editions de l'Universite de Bruxelle, Bruxelle.
43. Fitzwater, S., Hodes, Z. I. & Scheraga, H. A., 1978. Macromolecules 11: 805–811.
44. Blumenstein, M., Layne, P. P. & Najjar, V. A., 1979. Biochemistry 18: 5247–5253.
45. Sekacis, I. P., Liepins, E. E., Veretennikova, N. I. & Chipens, G. I., 1979. Bioorg. Chim. 5: 1617–1622.
46. Stabinsky, Y., Gottlieb, P., Zakuth, V., Spirer, Z. & Fridkin, M., 1978. Biochem. Biophys. Res. Commun. 83: 599–606.
47. Nair, R. M. G., Ponce, B. & Fudenberg, H. H., 1978. Immunochemistry 15: 901–907.
48. Bar-Shavit, Z., Stabinsky, Y., Fridkin, M. & Goldman, R., 1979. J. Cell. Physiol. 100: 55–62.
49. Bar-Shavit, Z., Bursker, I. & Goldman, R., 1980. Mol. Cell. Biochem., 30: 151–155.
50. Koren, H. S., Handwerger, B. S. & Wunderlich, J. R., 1975. J. Immunology 114: 894–897.
51. Constantopoulos, A., Likhite, V., Crosby, W. H. & Najjar, V. A., 1973. Cancer Res. 33: 1230–1234.
52. Constantopoulos, A. & Najjar, V. A., 1972. Cytobios. 6: 97–100.
53. Constantopoulos, A. & Najjar, V. A., 1973. J. Biol. Chem. 248: 3819–3822.
54. Erp, E. E. & Fahrney, D., 1975. Arch. Biochem. Biophys. 168: 1–7.
55. Bar-Shavit, Z., Goldman, R., Stabinsky, Y., Gottlieb, P., Fridkin, M., Teichberg, V. & Blumberg, S., 1980. Biochem. Biophys. Res. Commun. 94: 1445–1451.
56. Babior, B. M., 1978. N. Engl. J. Med. 298: 659–668.
57. Stossel, T. P., 1974. N. Engl. J. Med. 290: 717–723, 774–780, 833–839.
58. Baehner, R. L., Boxer, L. A. & Davis, J., 1976. Blood 48: 309–313.
59. Baehner, R. L. & Nathan, D. G., 1968. N. Engl. J. Med. 278: 971–976.

96

60. Spirer, Z., Zakuth, V., Golander, A., Bogair, N. & Fridkin, M., 1975. J. Clin. Invest. 55: 198–200.
61. Gottlieb, P., 1978. M.Sc. Thesis, The Weizmann Institute of Science, Rehovot, Israel, pp. 1–44.
62. Iguchi, H. & Nakazawa, S., 1976. Japan J. Bacteriol. 31: 81.
63. Stabinsky, Y., 1979. Ph.D. Thesis. The Weizmann Institute of Science, Rehovot, Israel, pp. 1–140.
64. Unanue, E. R., 1972. Advances in Immunology (Dioxon, F. J. & Kunkel, H. G., eds.), Vol. 15, pp. 95–165, Academic press, New York.
65. Ishizaka, K. & Adachi, T., 1976. J. Immunol. 117: 40–47.
66. Thomas, D. W. & Shevach, E. M., 1976. J. Exp. Med. 144: 1263–1273.
67. Thomas, D. W. & Shevach, E. M., 1977. Proc. Natl. Acad. Sci. USA 74: 2104–2108.
68. Tzehoval, E., Segal, S., Stabinsky, Y., Fridkin, M., Spirer, Z. & Feldman, M., 1978. Proc. Natl. Acad. Sci. USA 75: 3400–3404.
69. Tzehoval, E., Segal, S., Stabinsky, Y., Fridkin, M., Spirer, Z. & Feldman, M., 1979. Springer Semin. Immunopathol. 2: 205–214.
70. Tzehoval, E., Segal, S., Stabinsky, Y., Fridkin, M., Spirer, Z. & Feldman, M., 1979. Advances in Pharmacology and Therapeutics (Vargaftig, B. B., ed.), Vol. 4, pp. 137–144, Pergamon Press, Oxford and New York.
71. Florentin, I., Bruley-Rosset, M., Kiger, N., Imbach, J. L., Winternitz, F. & Mathe, G., 1978. Cancer Immunol. Immunother. 5: 211–216.
72. Nishioka, K., 1979. Br. J. Cancer 39: 342–345.
73. Noyes, R. D., Babcock, G. F. & Nishioka, K., 1980. Proceedings Amer. Assoc. Cancer Research and Amer. Soc. Clin. Oncol. 21, Abst. 1046.
74. Horsmanheimo, A., Horsmanheimo, M. & Fudenberg, H. H., 1978. Clin. Immunol. Immunopathol. 11: 251–255.
75. Nishioka, K., 1978. Gann. 69: 569–572.
76. Goetzl, E. J., 1976. Amer. J. Pathol. 85: 419–436.
77. Newman, W., Bloom, B. R. & Satoh, P., 1976. Cell Immunol. 27: 343.
78. Martinez, J., 1976. These Docteur de Science Physiques, Academie de Montpellier, pp. 1–144, Université des Sciences Techniques de Languedoc, Montpellier.
79 Braun, T., Hechter, O. & Rudinger, J., 1969. Endocrinology 85: 1092–1096.
80. Bonne, D. & Cohen, P., 1975. Peptides: Chemistry, Structure and Biology, Proceedings of the 4th American Peptide Symposium (Walter, R. & Meinhofer, J., eds.), pp. 711–717, Ann Arbor Science Publishers, Ann Arbor, Michigan.
81. Makino, T., Carraway, R., Leeman, S. E. & Greep, R. O., 1973. Biol. of Reproduction 9: 66.
82. Carraway, R. & Leeman, S. E., 1973. J. Biol. Chem. 248: 6854–6861.
83. Chang, D., Griebrokk, T., Knudsen, R., Howard, G., Humphries, J., Folkers, K. & Bowers, C. Y., 1975. Biochem. Biophys. Res. Commun. 65: 1208–1213.
84. Lembeck, F., 1953. Naunym Schmiedebergs Arch. Pharmakol. 219: 197–213.
85. Otsuka, M. & Konishi, S., 1977. Substance P (von Euler, U. S. & Pernow, B., eds.), pp. 207–214, Raven Press, New York.
86. Jessell, T. M. & Iversen, L. L., 1977. Nature 268: 549–551.
87. Henderson, L. E., Gilden, R. V. & Oroszlan, S., 1979. Science 203: 1346–1348.
88. Ogawa, Y., Quagliarotti, G., Jordan, J., Taylor, C. W., Starbuck, W. C. & Busch, H., 1969. J. Biol. Chem. 244: 4387–4392.
89. Oliveira, E. B., Gotschlich, E. C. & Liu, T-Y., 1977. Proc. Natl. Acad. Sci. USA 74: 3148–3151.
90. Oroszlan, S., Henderson, L. E., Stephenson, J. R., Copeland, T. D., Long, C. W., Ihle, J. N. & Gilden, R. V., 1978. Proc. Natl. Acad. Sci. USA 75: 1404–1408.
91. Luftig, R. B., Yoshinaka, Y. & Oroszlan, S., 1977. J. Cell Biol. 75: Abst. V1562.
92. Suk, W. A. & Long, C. W., 1979. Annual Meeting 1979, Amer. Soc. Microbiol. Abst. S105.
93. Eraklis, A. J., Kevy, S. V., Diamond, L. K. & Gross, R., 1967. N. Engl. J. Med. 276: 1225–1229.
94. Erickson, W. D., Burgert, E. O. & Lynn, H. B., 1968. Amer. J. Dis. Child. 116: 1–2.
95. King, H. & Shumaker, H. B., Jr., 1952. Amer. Surg. 136: 239–242.
96. Grinblat, J. & Gilboa, Y., 1975. Amer. J. Med. Sci. 270: 523–524.
97. Hyslop, N. E., Jr., 1975. N. Engl. J. Med. 293: 547–553.
98. Winter, S. T., 1974. Clin. Pediatr. 13: 1011–1012.
99. Constantopoulos, A., Najjar, V. A. & Smith, J. W., 1972. J. Pediat. 80: 564–572.
99a. Najjar, V. A., 1975. J. Pediat. 87: 1121–1124.
100. Spirer, Z., Zakuth, V., Diamant, S., Mondorf, W., Stefanescu, T., Stabinsky, Y. & Fridkin, M., 1977. Brit. Med. J. 2: 1574–1576.
101. Sheagren, J. N., Block, J. B. & Wolff, S. M., 1967. J. Clin. Invest. 46: 855–862.
102. Stabinsky, Y., Bar-Shavit, Z., Fridkin, M. & Goldman, R., 1980. Mol. Cell. Biochem. 30: 71–77.
103. Rivkin, I., Rosenblatt, J. & Becker, E. I., 1975. J. Immunol. 115: 1126–1134.
104. Hill, H. R., 1978. Leukocytes Chemotaxis: Methods, Physiology and Clinical Implications (Gallin, J. I. & Quie, P. G., eds.), pp. 179–193, Raven Press, New York.
105. Spirer, Z., Zakuth, V., Golander, A. & Bogair, N., 1975. Experientia 31: 118–119.
106. Zurier, R. B., Weissmann, G., Hoffstein, S., Kammerman, S. & Tai, H. H., 1974. J. Clin. Invest. 53: 297–309.
107. Ignaro, L. J. & George, W. J., 1974. Proc. Natl. Acad. Sci. USA 71: 2027–2031.
108. Estensen, R. D., Hill, H. R., Quie, P. G., Hogan, N. & Goldberg, N. A., 1973. Nature 245: 458–460.
109. Sandler, J. A., Gallin, J. I. & Vaughan, M., 1975. J. Cell. Biol. 67: 480–484.
110. Hill, H. R., Estensen, R. D., Quie, P. G., Hogan, N. A. & Goldberg, N. D., 1975. Metabolism 24: 447–456.
111. Goldberg, N. D. & Haddox, M. K., 1977. Ann. Rev. Biochem. 46: 823–896.

112. Schultz, G., Hardman, J. G., Schultz, K., Baird, C. E. & Sutherland, E. W., 1973. Proc. Natl. Acad. Sci. USA 70: 3889–3893.

113. Smith, R. J. & Ignarro, L. J., 1975. Proc. Natl. Acad. Sci. USA 72: 108–112.

114. Ferrendelli, J. A., Kinscherf, D. A. & Chang, M. M., 1973. Mol. Pharmacol. 9: 445–454.

115. Manganiello, V. & Vaughan, M., 1972. Proc. Natl. Acad. Sci. USA 69: 269–273.

116. Catane, R., Schlanger, S., Gottlieb, P., Halpern, J., Treves, A. J., Fuks, Z. & Fridkin, M., 1981. Submitted to the American Society of Clinical Oncology Meeting.

Received August 4, 1980.

Tuftsin analogs and their biological activity

Ignacy Z. Siemion and Danuta Konopinska
Institute of Chemistry, Wroclaw University, 50–383 Wroclaw, Poland

Summary

In this paper the literature data on the structure-activity relationship for the series of tuftsin analogs are summarized. Among others, the questions of the substitution of particular amino acid residues in different positions of the peptide chain, as well as the questions of shortening and lengthening of the peptide chain of tuftsin, are reviewed. The existing models of the biologically active conformation of tuftsin are also summarized.

Introduction

The naturally occcurring tetrapeptide tuftsin, Thr-Lys-Pro-Arg, was discovered and isolated by Najjar et al. (1, 2, 3) in 1973 on the basis of its ability to stimulate the phagocytic activity of polymorphonuclear granylocytes. It is found in a leukophilic γ-globulin fraction of human blood, where it is covalently bonded to its carrier molecule leukokinin (4). In a similar fraction of canine blood serum, the following tuftsin, analog, Thr-Lys-Pro-Lys, was recently discovered (4). From human leukokinin obtained from a patient with tuftsin deficiency and which was deprived of phagocytic activity, the tetrapeptide Thr-Glu-Pro-Arg was isolated. It had no phagocytosis-stimulating activity (4).

According to a mechanism proposed by Najjar (4, 5) tuftsin is liberated by the action of two enzymes. The first tuftsin endocarboxypeptidase, splits the Arg-Glu peptide bond at the carboxyterminal residue. This process takes place in the spleen where leukokinin is transported in the blood. The second enzyme, leukokininase, which is present on the outer surface of phagocytic cell membrane, splits the amino-terminal Lys-Thr peptide bond liberating tuftsin.

The first chemical synthesis of tuftsin was performed by Najjar et al. (3) by means of the solid-phase procedure. At the same time, three tuftsin analogs: Thr-Lys-Pro, Lys-Pro-Arg, and Thr-Lys-Pro-Pro-Arg were synthesized (3); Lys-Pro-Arg, Thr-Lys-Pro-Pro-Arg, was well as tuftsin retro-enantiomer D-Arg-D-Pro-D-Lys-D-Thr (6) were found to be phagocytosis inhibitors (3, 6, 7). Thereafter tuftsin was synthesized in several laboratories by classical methods (8–14). Some examples of these syntheses are shown on Schemes 1–3.

Recently, the broad spectrum of the biological activities of tuftsin has been reviewed by Najjar (4, 7, 15). Besides phagocytosis stimulation, other activities like stimulation of motility of granulocytes, influence on antibody formation, promotion of bacterial killing properties and tumoricidal activity of phagocytic cells, as well as augmentation of immunogenic function of macrophages were found to be stimulated by tuftsin. Very recently we discovered (16) that after administration of tuftsin into the rat brain analgesia is produced. This long lasting (about 30 min) effect was not influenced by naloxone. This suggests that the molecular mecha-

Molecular and Cellular Biochemistry 41, 99–112 (1981).   0300-8177/81/0041-0099/$02.80.

100

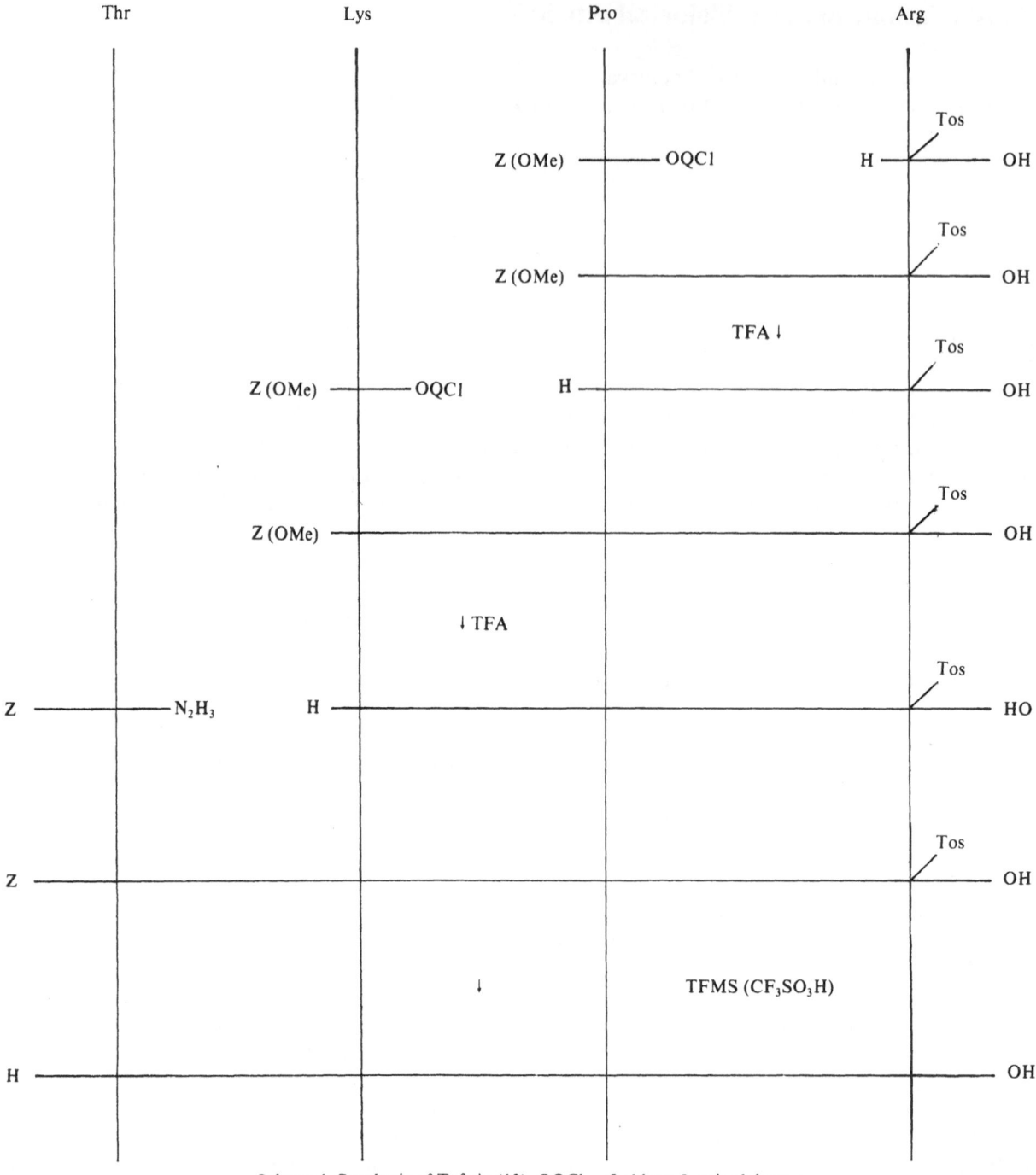

Scheme 1. Synthesis of Tuftsin (13). OQCl = 5-chloro-8-quinolyl ester.

101

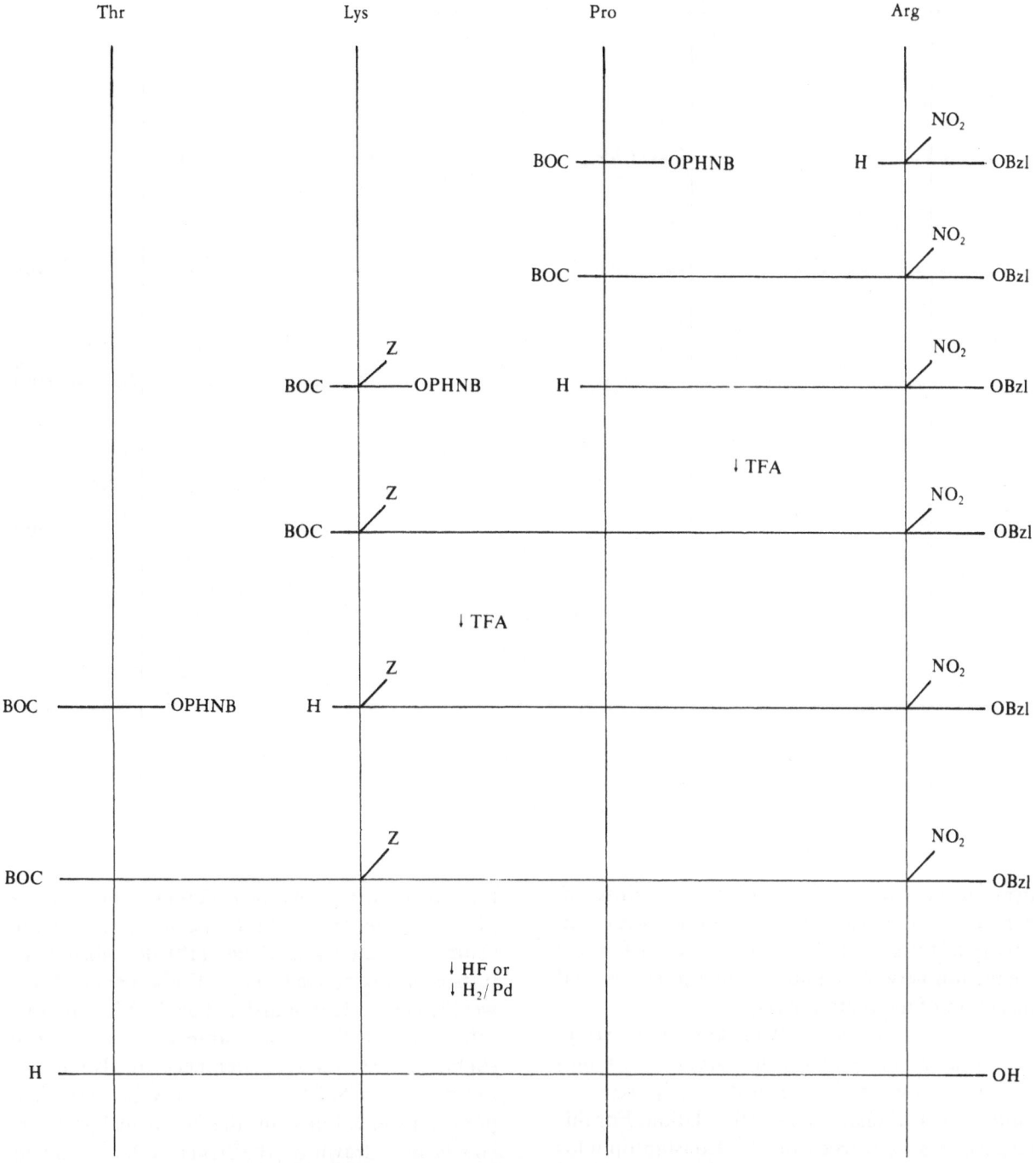

Scheme 2. PHNB – (4-hydroxy-3 nitro)-benzylated polystyrene. Synthesis of tuftsin (11, 12).

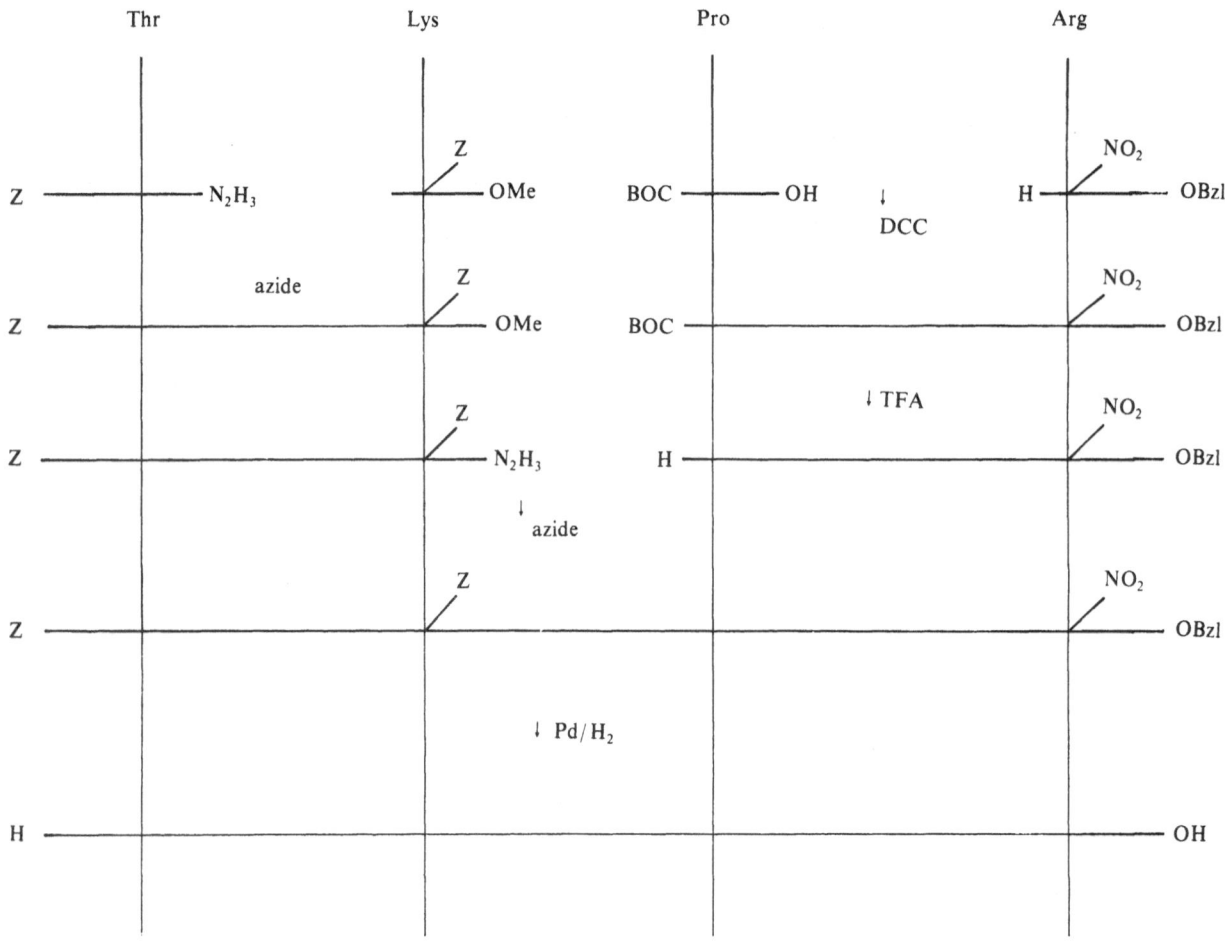

Scheme 3. Synthesis of tuftsin (8, 9).

nisms of the analgesic action of tuftsin is different from that of the enkaphalins. From this unexpected activity of tuftsin it follows that there exists some connection between immunological processes and the action of the central nervous system.

Since the discovery of tuftsin about eighty structural analogs have been synthesized. Their biological activity was investigated mainly from the point of view of phagocytosis stimulation. For this purpose, assays worked out by Constantopoulos and Najjar (7) or Fridkin et al. (12) are used. The latter authors (12) also used the ability of tuftsin to stimulate the reduction of nitroblue tetrazolium after phagocytosis of the dye-haparin or dye-fibrin complex. This, too, was used for testing the biological activity of tuftsin analogs. For quantitative determination of tuftsin in blood, a radio-immunoassay was elaborated by Spirer et al. (18).

It should be pointed out that in several cases for the same tuftsin analog different values of the phagocytic index can be found in literature. For example Koenig and Seiler (19) described Ser[1]-tuftsin as very active whereas Fridkin et al. (12) as weakly active. Konopinska et al. (20) found Gly[3]-tuftsin as inactive, but Stabinsky et al. (21) as slightly active. Similar differences are also quoted fo Val[1]-tuftsin (8, 12). Consequently, because of the poor reproducibility of the biological test, the conclusions drawn on the structure-biological activity relationship for tuftsin analogs have to be considered with caution.

Dependence of tuftsin phagocytic activity on the length of the peptide chain

According to the literature (2, 12) the tripeptide fragments of tuftsin, such as Lys-Pro-Arg and Thr-

Lys-Pro are devoid of phagocytosis-stimulating activity. However, for Lys-Pro-Arg, being the C-terminal tripeptide, a low inhibitor activity was found. On the other hand Stabinsky et al. (21) found recently a low phagocytosis-stimulating activity for des Pro³-tuftsin (Thr-Lys-Arg).

The amino acid sequence of tuftsin appears in the C_H2 domain of human H chain of γ-immunoglobulin EU, forming the fragment 289–292 (22). However, synthetic oligopeptides elongated on their N-terminus by vicinal amino acid residues of the tuftsin sequence of the peptide chain: Lys-Thr-Lys-Pro-Arg and Ala-Lys-Thr-Lys-Pro-Arg, were found to be biologically inactive (23). Extension of the tuftsin peptide chain on its N-terminus by the N-succinimidyl-3-(4-hydroxy-5-I¹²⁵-phenyl)-propionate moiety gives a product with low binding ability to polymorphonuclear leukocytes (24). A synthetic pentapeptide Thr-Gly-Lys-Pro-Arg was found to be inactive (25). On the other hand the pentapeptide tuftsinyl-glycine (tuftsin-Gly⁵) possesses about 40% activity of tuftsin itself (21). Tuftsinyl (I¹²⁵)-tyrosine (tuftsin-Tyr⁵) also binds well to tuftsin receptors on the membranes of polymorphonuclear leukocytes (24). Finally, extension of the tuftsin peptide chain by insertion of an additional proline residue, giving the Thr-Lys-Pro-Pro-Arg pentapeptide, leads to a product with strong inhibitory activity (3).

An octapeptide, tuftsinyl-tuftsin which was synthesized recently (4) possesses little or no phagocytic activity; it acts, however, as a very potent tumoricidal factor.

Based on these data it can be concluded that tuftsin is the optimum tetrapeptide unit. It is evident, however, that shortening or extending of the peptide chain on its C-terminus gives, from case to case, peptides that show distinct binding to leukocytes. This suggests that the C-terminus of the tuftsin peptide chain is less sensitive to chemical modifications that the N-terminus.

The problem of biologically active conformation of tuftsin

During peptide-receptor interaction, a specific biologically active conformation of the peptide is generated. This biologically active conformation can differ from the peptide conformation observed in the solid state, or from solution conformation. In solution, many different peptide conformations exist in a state of conformational equilibrium. The picture which can be derived from NMR or CD spectroscopic measurements yields a mean conformation of the given peptide. However, it seems reasonable to assume that the conformation that is privileged in solution may be close to that of the biologically active conformation.

For the creation of a biologically active confor-

Table 1. Tuftsin analogs with shortened and prolongated amino acid residues.

Peptides	Biological activity relative to tuftsin	Inhibitory	Reference
Thr-Lys-Arg	0.2	not done	21
Lys-Pro-Arg	0.0	+	11
Thr-Lys-Pro	0.0	0	2, 3, 7
Thr-Lys-Pro-Pro-Arg	0.0	+	2, 3, 7, 50
Thr-Gly-Lys-Pro-Arg	0.0	not done	25
Lys-Thr-Lys-Pro-Arg	0.0	not done	23
Ala-Lys-Thr-Lys-Pro-Arg	0.0	not done	23
Tyr-Thr-Lys-Pro-Arg	0.0	+	11, 12
Acetyl-Thr-Lys-Pro-Arg	0.0	+	11, 12
(H_2N-phenylacetyl)-Gly₂-Thr-Lys-Pro-Arg	0.0	not done	4
Thr-Lys-Pro-Arg-Gly	0.5	not done	21
Thr-Lys-Pro-Arg-NH₂	0.6	not done	21
Thr-Lys-Pro-Arg-Thr-Lys-Pro-Arg	tumoricidal (probably)		4

Tuftsin activity as 1.0.

104

mation of tuftsin, the presence of a proline residue in position *3* of the peptide chain may be of great importance because of the spatial requirements introduced by the rigid pyrolidine ring of proline. However, NMR studies performed with linear proline containing oligopeptides showed that in solution several conformations exist at equilibrium (26). This result coincides with predictions derived from theorotecial calculations (27).

Few years ago (8) we proposed a β-turn conformation for biologically active tuftsin. This can be stabilized by a $4 \leftarrow 1$ intramolecular hydrogen bond between Thr^1 and Arg^4 residues and by ionic interaction between the terminal amino- and carboxy-group. Inspection of molecular models suggested that this tuftsin conformation may be of type III (28), for which values of dihedral angles $\Phi \text{ Lys}^2 = \Phi \text{ Pro}^3 = -60°$ and $\psi \text{ Lys}^2 = \psi \text{ Pro}^3 = -30°$ are expected (29). The scheme of this structure is represented in Fig. 1. In this type of β-turn, the Pro^3-residue is situated in a manner similar to the proline residue in gramicidin S (30). From data on β-turns appearing in globular proteins (31), it is known, however, that the proline residue is very seldom found in position (i + 2) of the bend, which has to be expected in the case of tuftsin. In globular proteins, if in this position of the turn the proline residue appears, then the X-Pro amide bond frequently exists in cis-configuration, creating the turn of type VI. From theoretical calculations (27) it also follows that there is a preference of proline to be in position (i + 1) instead (i + 2) of the turn. This

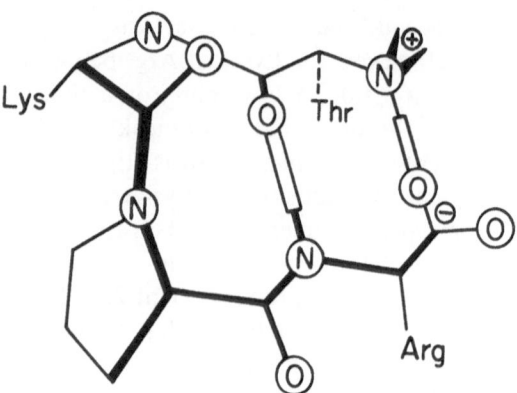

Fig. 1.

preference disappears, however, when hydration energy is taken into account (32) in the calculation.

In favour of our conformational hypothesis the following arguments could be brought out:

1. Investigation of IR spectra of chloroform and tetrachloromethane solutions of protected tuftsin analogs (33) showed that all these compounds form folded conformations with intramolecular hydrogen bonds. In the spectra of analogs possessing Pro residue in position 3 in the region of amide I absorption band, a picture characteristic for β-turn conformation was observed. The formation of β-turn should increase for analogs with D-amino acid residues (34). This was really the case for tuftsin analogs containing D-amino acid residues excluding D-Arg^4-tuftsin (35).

2. CD spectra of tuftsin (14) and its analogs (36)

Table 2. The biological activity of tuftsin analogs modified in position 1.

Peptides	Biological activity relative to tuftsin	Inhibitory	Reference
Tuftsin (Thr-Lys-Pro-Arg)	1.0	–	2,3
Leu-Lys-Pro-Arg	1.0	not done	47
D-Ser-Lys-Pro-Arg	1.0	not done	19
Pyro-Glu-Lys-Pro-Arg	1.0	not done	19
Ser-Lys-Pro-Arg	1.0	not done	19
Ser-Lys-Pro-Arg	0.0–0.2	+	11, 12
Lys-Lys-Pro-Arg	0.0–0.3	0	11, 12
Ala-Lys-Pro-Arg	0.0	+	11, 12
Val-Lys-Pro-Arg	0.0	+	11, 12
Val-Lys-Pro-Arg	0.5	not done	8
Acetyl-tuftsin	0.0	+	11, 12
O = C $\overset{\frown}{\smile}$ Thr-Lys-Pro-Arg	0.5	not done	45
D-Thr-Lys-Pro-Arg	slightly active	not done	28
D-Leu-Lys-Pro-Arg	0.0	not done	28

Tuftsin activity as 1.0.

are pH dependent. This observation suggests that their comformation in water is influenced by ionic interactions between the charged groups. Analysis of pH induced changes (36) suggests that they are probably evoked by an increase of the amount of β-turn conformers in neutral water solution. With L-amino acid residues no cis-Pro signals were observed (36). This indicates a reduced conformational lability of tuftsin in comparison to other linear proline containing oligopeptides (37, 38). If, however, position *3* of the tuftsin is occupied by a D-Pro residue the amounts of cis X-Pro and trans X-Pro forms of the molecules is practically equal. For tuftsin in the zwitterion form, the chemical shifts of C^βPro and C^βLys are consistent with β-turn conformation (36).

3. D-Lys^2-tuftsin, for which the β-turn structure was confirmed by CD measurements (36), possessses 10–20% of tuftsin phagocytic activity (28). However, there are some substantial arguments against the above mentioned hypothesis:

1. Very recently Blumenstein et al. (39) and independently Sekacis et al. (40) presented the results of ^{1}H-NMR and ^{13}C-NMR measurements of tuftsin. In contrast to CD data, relaxation time (NT$_1$) measurements in water do not confirm the existence of ionic interactions between the charged groups, which could stabilize a definite conformation of the peptide chain. Both research groups found, however, in DMSO solution a particular tuftsin conformation where Arg^4-amide proton is shielded from the solvent. According to Blumenstein et al. (39) some correlation can exist between biologically active, and DMSO tuftsin conformation. In the opinion of these authors DMSO conformation of tuftsin is not of the β-turn type, because the observed value of J_{NHCH} coupling constant 7.5 Hz of Lys residue is too large. It has to be pointed out, however, that a relatively small increase of Φ Lys angle from –60° to –80° is followed by an increase of J_{NHCH} coupling constant from 3–4 Hz to about 7 Hz (41). According to Sekacis et al. (40) the DMSO conformation of tuftsin may be of β-turn type.

2. Blumenstein and Najjar (39) pointed out that the β-turn conformation proposed by us is difficult to reconcile with the ability of Thr-Lys-Pro-Pro-Arg pentapeptide to bind with the tuftsin-receptors producing an inhibitory effect.

3. We have found (28) that D-Arg^4-tuftsin, which demonstrates, in its protected form, a low folding tendency (35) possesses about 40% of the phagocytic activity of tuftsin. On the other hand D-Val^3-tuftsin and D-Leu^1-tuftsin are biologically inactive (28), although the spectra of their protected derivatives indicate the presence of β-turn structure.

In 1978 Nikiforovich (42) proposed a different, low energy, conformation of tuftsin (see Fig. 2). According to this proposition the tuftsin peptide chain is folded forming a quasi-cyclic structure, stabilized by ionic interaction between the carboxylic group of Arg^4 and the ϵ-amino group of Lys^2. Such a conformation is described by following dihedral angles: Φ Lys = –125°, ψ Lys = +120° and ψ Pro = –50°. A similar structure was proposed by Vicar et al. (14), based on the pH dependence of CD curves. As was noted above, the changes in CD curves however can be also be interpreted in favour of the β-turn conformation in neutral water solution.

Recent NMR measurements of Sekacis et al. (40) are against the Nikiforovich's model of a privileged conformation of tuftsin. Comparison of the data of relaxation times determination performed for tuftsin and independently for its N$^\epsilon$-Lys^2-benzyloxy-

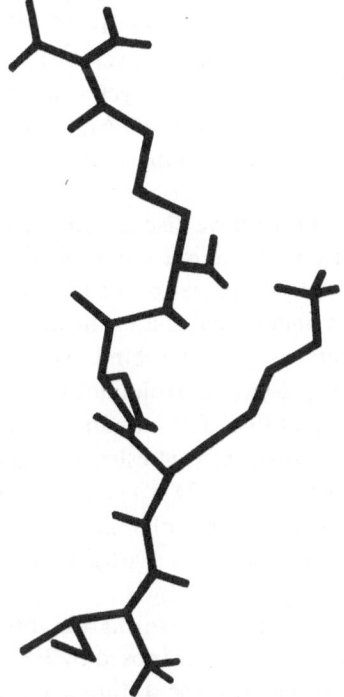

Fig. 2.

106

Table 3. Tuftsin analogs modified in position 2.

Peptides	Biological activity relative to tuftsin	Reference
Thr-Arg-Pro-Arg	1.0	44
Thr-Orn-Pro-Arg	0.0–0.5	25, 43
Thr-Leu-Pro-Arg	0.0	23
Thr-D-Lys-Pro-Arg	0.1–0.2	28
Thr-Glu-Pro-Arg	0.0	4

Tuftsin activity as 1.0.

carbonyl derivative showed that the ionic interaction postulated for Nikiforovich's quasi-cyclic structure of tuftsin does not exist in fact. Further arguments against this model results from the determination of biological activity of some tuftsin analogs. Substitution of Lys[2] by ornithine (43) or arginine (44) leads to active tuftsin analogs, although the spatial relations between the amino acid side chains in these analogs should be different from those in tuftsin. Also the substitution of Lys[2] by leucine (23) and alanine (9) gives active analogs although the formation of Nikiforovich's structure is here impossible. This structure should also be sensitive to configurational changes in position *2* of the peptide chain. In this connection, D-Lys[2]-tuftsin possesses 10–20% of the biological activity of tuftsin itself (28). In favour of Nikiforovich's conformational model, however, is the observation of Chipens et al. (53) that cyclo-tuftsin, a peptide with covalent bridge between Arg[4]-carboxyl group and N[ε] Lys[2]-amino group, exhibits phagocytosis stimulating activity.

In both of the above discussed models of tuftsin conformation, the important role of ionic interactions for the stabilization of the proper spatial structure is underlined. It must be noted, however, that N-acetyl-tuftsin can bind with the tuftsin-receptor (12); 2-oxo-oxazolidine derivative of tuftsin, with modified Thr[1]-residue, is biologically active (45); tuftsin methyl ester also binds well to human neutrophiles (24). These data render the thesis on the importance of ionic interactions in the creation of biologically active tuftsin conformation doubtful.

A model of tuftsin conformation different from both of the above was proposed by Fitzwater et al. (46). According to these authors, the global minimum area of tuftsin energy surface is characterized by many individual conformations, but all low-energy conformers have the same general structure, described as a 'hairpin with two split ends'. Both trans Lys-Pro and cis Lys-Pro units are possible in this structure, but the ensemble is dominated by trans-conformers. In the trans-conformers, the outer portions of the splits are formed by the peptide backbone, and the inner portions by the Lys[2] and Arg[4] side chains. For cis-conformers the situation is the opposite. The values of ψ Lys angle are restricted to two narrow bands between ψ +80–90°, and ψ +150–160°; Lys-Pro fragments of the molecule occurs in only a few conformations (DA, DC, DF, EA, EC, EF, FC, and AC), while the conformation of Thr[1] and Arg[4] are less restricted. In the peptide conformation which corresponds to global energy minimum, the conformation of Pro-residue is of the C-type. In this conformation an intramolecular hydrogen bond between C = O Lys[2] and NH of Arg[4] can exist. (Fig. 3)

In the opinion of Fitzwater et al. (46) the action

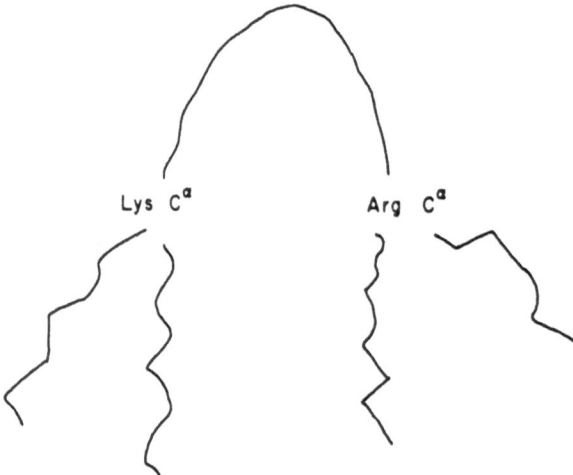

Fig. 3.

Table 4. Tuftsin analogs modified in the position 4.

Peptides	Biological activity relative to tuftsin	Inhibitory	Reference
Thr-Lys-Pro-Lys	0.5–0.6	not done	1, 43, 44
Thr-Lys-Pro-Arg-NH$_2$	0.6	not done	21
Thr-Lys-Pro-Ala	0.7	not done	8, 9
Thr-Lys-Pro-Homo-Arg	1.0	not done	21
Thr-Lys-Pro-Nor-Arg	0.5	not done	21
Thr-Lys-Pro-N^{G}Me-Arg	0.5	not done	21
Thr-Lys-Pro-His	0.3	not done	21
Thr-Lys-Pro-Orn	0.0	not done	25
Thr-Lys-Pro-Arg-NO$_2$	0.0	0	11, 12
Thr-Lys-Pro-D-Arg	04.	not done	21, 28
Thr-Lys-Pro-Leu	0.2	not done	49

Tuftsin activity as 1.0.

of tuftsin does not depend upon a narrowly defined molecular conformation. The charge on the side chains of Lys^2 and Arg^4 appears to have only a minor influence on the phagocytic activity, and the major factor influencing activity is a bulky side chain in position 3.

It must be remarked that the preference for all-trans conformers of tuftsin predicted by Fitzwater et al. (46) finds support in experimental data (36, 38, 40). It is also worth recalling that ^{1}H-NMR data (39, 40) indicate the participation of NH of Arg^4 in intramolecular hydrogen bonding. The analysis of ^{13}C-NMR data (36, 40) however, shows that the C-type conformation of Pro required by the model of tuftsin does not appear in tuftsin solutions. Thus, in the light of these data the model of tuftsin conformation proposed by Fitzwater et al. (46) also seems to be rather questionable.

Biological significance of particular amino acid residues of tuftsin

Thr^1-residue

As it was noted above, extending the peptide chain of tuftsin on its N-terminus leads to a distinct decrease in biological activity. From this, it can be concluded that the N-terminal part of the tuftsin molecule is essential for its biological effects. This conclusion corresponds well with the observation of Fridkin et al. (12) that N-acetyl-tuftin and also p-aminophenylacetyl-tuftsin have only a low in-

hibitory effect in experiments with heat-killed yeast cells. However some tuftsin analogs with protected terminal amino group are biologically active; active are Pyro-Glu^1-tuftsin (19) and 2-oxo-oxazolidine-tuftsin (45), although the basic properties of the terminal amino group in both of these derivatives are strongly reduced. It cannot be excluded, however, that under the conditions of the biological tests, both these derivatives are metabolized to peptides with a free terminal amino group.

In the opinion of Stabinsky et al. (21) the integrity of the N-terminal threonine residue of tuftsin is of great importance for the preservation of biological action of the peptide. Indeed, substitution of Thr^1 by other amino acids leads, as a rule, to analogs with reduced biological activity or inhibitory properties. According to Fridkin et al. (12) Val^1-and Ala^1-tuftsin exhibit an inhibitory effect to tuftsin's action, whereas Lys^1-and Ser^1-tuftsin are less active than tuftsin itself. The last analog shows also inhibitory activity. Ala^1-tuftsin inhibits not only phagocytosis stimulation, but also represses nitroblue tetrazolium reduction. This effect was not observed for other analogs modified in position *1*. On the other hand Matsuura et al. (47), using Consantopoulos and Najjar's (17) test, found Leu^1-tuftsin even more active than tuftsin itself. Koenig and Seiler (19) described Ser^1- and also D-Ser^1-tuftsin as active compounds. Konopinska et al. (8) found also that Val^1-tuftsin stimulates phagocytosis. Investigating the influence of configurational changes on biological activity of tuftsin, Nawrocka et al. (28) detected a small activity for D-Thr^1-

108

Table 5. Tuftsin analogs modified in position 3.

Peptide	Relative biological activity to tuftsin	Origin	Reference
Arg³-tuftsin	0.9	transversional, one point mutation	20, 48
			20, 48
Ala³-tuftsin	0.6	idem	
Leu³-tuftsin	0.5	transitional, one point mutation	20, 48
			20, 48
Ser³-tuftsin	0.0	idem	
Thr³-tuftsin	0.9	C → A exchange, succes-	20, 48
Gln³-tuftsin		sive transition-trans-	20, 48
His³-tuftsin	0.0	version or transversion-transition	20
Lys³-tuftsin	0.5	control group	20
Ile³-tuftsin	0.4	control group	20
Val³-tuftsin	0.5	control group	20, 48
Tyr³-tuftsin	0.3	control group	20
Phe³-tuftsin	0.0	control group	20
Asp³-tuftsin	0.0	control group	20
Sarc³-tuftsin	0.0	control group	20
Gly³-tuftsin	0.0	control group	20
D-Ala³-tuftsin	0.0	control group	28
D-Leu³-tuftsin	0.0	control group	28
D-Arg³-tuftsin	0.0	control group	28
D-Thr³-tuftsin	slightly active	control group	28
D-Pro³-tuftsin	slightly active	control group	28

Tuftsin activity as 1.0.

tuftsin, whereas D-Leu[1] analog was inactive.

The above observations lead to the conclusion that a terminal amino group could be of importance for the generation of a biological effect of tuftsin. It seems, however, that the side chain of the amino acid in position *1* of tuftsin is more resistant to structural changes than was postulated by Stabinsky et al. (45).

Basic amino acids: Lys² and Arg⁴

In 1973 Constantopoulos and Najjar (17) postulated that basic amino acids are responsible for the interaction of tuftsin with its receptor. This idea was based on the observation that removal of sialic acid from the surface of granulocyte cells destroys the sensitivity of the cell to tuftsin. In general, this idea was confirmed by further investigations. In 1977 Fridkin et al. (12) showed that ω-nitro⁴-tuftsin is devoid of phagocytosis-stimulating and also possess inhibitory activity. The importance of basic residues for the generation of biological effects of tuftsin was also confirmed by experiments with analogs in which consecutive positions of the peptide chain were replaced by leucine. As stated above, Leu[1]-tuftsin is more active than tuftsin itself (47). Markedly active is also Leu³-tuftsin (48). On the other hand Leu²-tuftsin (23) and Leu⁴-tuftsin possess only very weak activity. These observations indicate that positions *1* and *3* of the peptide are less sensitive to substitution by leucine than position *2* and *4*. Less distinct results were obtained by the substitution of the basic residues with alanine. For Thr-Ala-Val-Arg, Thr-Ala-Arg-Lys and Thr-Lys-Pro-Ala a pronounced stimulatory effect was found (8, 9).

On the other hand, Lys[1]-tuftsin, an analog with additional positive charge, is much less active than tuftsin itself (12). No stimulatory or inhibitory activity was observed also for lysine oligomers containing 3, 4, 6 and 20 lysine residues (12).

As it was shown by Martinez et al. (10) the interchange of basic Lys² and Arg⁴ residues in the tuftsin molecule leads to an active analog. Also Hisatsune et al. (44) showed that both basic residues of tuftsin can be exchanged without loss of activity: Thr-Arg-Pro-Arg; Thr-Lys-Pro-Lys, and Thr-Arg-Pro-Lys are, according to these authors,

Table 6. Tuftsin analogs modified in positions 2 and 4, and 1 and 4.

Peptides	Biological activity relative to tuftsin	Reference
Thr-Orn-Pro-Ala	0.4	43
Thr-Arg-Pro-Lys	1.0	44
O=C ⟲Thr-Lys-Pro-Arg-NH$_2$	0.3	21
O=C ⟲Thr-Lys-Pro-Nor-Arg	0.1	21
O=C ⟲Thr-Lys-Pro-His	0.1	21

Tuftsin activity as 1.0.

Table 7. Other tuftsin analogs.

Peptides	Biological activity relative to tuftsin	Reference
Arg-Pro-Lys-Thr	1.0	25
Arg-Pro-Lys-Thr	0.0	44
(Lys)$_n$ n = 3, 4, 6, 20	0.0	11, 12
Lys-Thr-Pro-Arg	active	51
Lys-Thr-Arg-Pro	not done	51
Thr-Pro-Lys-Arg	not done	51
Thr-Lys-Arg-Pro	not done	51
Thr-Pro-Arg-Lys	not done	52

Tuftsin activity as 1.0.

comparable in their stimulatory potency with tuftsin itself. Lys[4]-tuftsin was also synthesized by Konopinska et al. (43) and yielded 50% of tuftsin activity. The same authors found (43) that also Thr-Lys-Arg-Arg tetrapeptide is a very potent tuftsin analog.

The dependence of biological activity on the position of basic residues in the peptide chain of tuftsin is not fully clear. Thus, Hisatsune et al. (44) stated that retro-tuftsin (Arg-Pro-Lys-Thr) with basic residues in positions *1* and *3* is devoid of activity. This could be evoked, however, by the wrong localization of Pro, causing a change of the spatial conformation of the peptide. The lack of biological activity was also observed (43) for Thr-Pro-Lys-Ala tetrapeptide in which Pro-residue occupies the position *2*. However, in contrast to the finding of an inactive retro-tuftsin (44) Yasumura et al. (25) reported high activity similar to that of tuftsin itself.

A mutual replacement of Thr[1] and Lys[2] in the tuftsin molecule leads to a very active phagocytosis-stimulating analog (50), which evokes also chemotaxis of PMN (51). Biological activity was found for Leu-Lys-Lys-Ala (43) and Thr-Lys-Lys-Ala (9), whereas Thr-Gly-Gly-Lys and Thr-Lys-Ala-Ala were inactive (9). For the synthetically prepared tetrapeptides: Thr-Lys-Arg-Pro, Lys-Thr-Arg-Pro, and Thr-Pro-Lys-Arg (50) the activity was not specified. It is worh noting, however, that a tetrapeptide with very similar amino acid sequence, Thr-Pro-Arg-Lys, was found to be a contraceptive (52).

Extending or shortening of the side chains of basic amino acids of tuftsin distinctly modifies the activity of corresponding analogs. It was found (43) that Thr-Orn-Pro-Arg and Thr-Orn-Pro-Ala have 50% and 30% of tuftsin activity respectively. Very recently Stabinky et al. (21) found that homo-Arg[4]-tuftsin is as active as tuftsin while nor-Arg[4]-tuftsin lost about 50% of phagocytic activity. Masking of the guanidine moiety of Arg[4] by methyl group such as in N[G]-methyl-Arg[4]-tuftsin and N[G]-methyl-homo-Arg[4]-tuftsin decreases phagocytic activity by 50%; for His[4]-tuftsin 25–30% of tuftsin activity was found.

The modification of tuftsin structure which involves not only the basic residues but also N-terminal Thr as in its transformation into 2-oxo-oxazolidine derivative, diminishes activity as compared with Lys[4]-tuftsin, nor-Arg[4]-tuftsin, and His[4]-tuftsin, respectively (21).

The change of configuration of basic residues causes also lowering of activity. For D-Lys[2]-tuftsin 10–20% and for D-Arg[4]-tuftsin about 40% tuftsin activity was found (28). The same activity was also found for D-Arg[4]-tuftsin by Stabinsky et al. (21).

From the observations presented above, it can be concluded that the basic amino acid residues are very important for the generation of the biological effects of tuftsin. Active analogs can however, be obtained by a broad range of structural modifications, which involve:

i. extending or shortening of side chain in basic residues; ii. replacing of basic residues within the peptide chain; iii. configurational changes involving basic residues; iv. substitutions of a definite basic residue by a nonpolar amino acid.

As a rule such modifications diminish the biological activity. It seems also that the presence of two basic residues is not necessary for biological activity. Also tetrapeptides containing three basic

Table 8. Fragments of protein as tuftsin analogs.

Peptides	From:	Biological activity relative to tuftsin	Reference
Thr-Lys-Pro-Ala	histones	0.7	8,9
Thr-Ala-Val Arg	histones	0.7	8,9
Thr-Ala-Arg-Lys	histones	0.6	8,9
thr-Gly-Gly-Lys	histones	0.1	8,9
Thr-Lys-Ala-Ala	histones	0.1	8,9
Thr-Lys-Lys-Ala	histones	0.7	8,9
Leu-Lys-Lys-Ala	histones	0.6	8,9
Thr-Pro-Lys-Ala	B-chain of insulin	0.0	8,9
Thr-Leu-Pro-Arg	light chain of immunoglobulin (AG)	0.0	23
Lys-Thr-Lys-Pro-Arg	heavy chain of immunoglobulin (EU)	0.0	23
Ala-Lys-Thr-Lys-Pro-Arg	heavy chain of immunoglobulin (EU)	0.0	23

Tuftsin Activity as 1.0.

amino acid residues were found to be active. In this connection, however, the lack of activity of tetra-Lys seems to be a very interesting phenomenon, which deserves attention.

Pro³-residue

Because of its rigid structure the proline moiety in position *3* of the peptide chain of tuftsin seems to be necessary for the biologically active conformation of the peptide. It was shown (20, 48), however, that substitution of Pro by other amino acids, leading to more flexible molecules preserves biological activity in many cases. The proline sequence is coded by the triplets: CCU, CCC, CCG and CCA. With this in mind, tuftsin analogs were synthesized in which the Pro³-residue was substituted by the amino acid residues resulting from a one-point mutation in the Pro codons. It was found that analogs resulting from transversional one-point mutations, Thr-Lys-Arg-Arg and Thr-Lys-Ala-Arg, are highly potent. From the two transitional one-point mutation analogs, Leu³-tuftsin has 50% activity, but Ser³-tuftsin is inactive. In the group of analogs that would result from replacing cytosine (C) with adenine (A) for the corresponding Pro codons, Thr³-tuftsin was found to be almost equal to tuftsin in its activity, whereas Gln³-tuftsin and His³-tuftsin showed less and lesser activity, respectively. The control group comprised analogs containing in position *3* such amino acid residues in

which codons differ from those of Pro by more than one base (Lys, Ile, Val, Tyr, Phe, Asp, and Gly); to this group sarcosine³-tuftsin was added. Most of these analogs possessed strongly reduced activity and half of them were almost inactive.

Based on these results a hypothesis on the existence of functional conservatism in the coding of tuftsin sequence was formulated (20, 48).

A comparison of the structures of the side chains of X³-analogs of tuftsin which possessed 40% or higher biological activity, showed that all of them except Ala³-tuftsin have a γ-methyl- or γ-methylene group. This suggested that the proline ring in tuftsin could be essential not only for the creation of the proper peptide conformation, but also for the interaction with the receptor by its own γ-methylene group. In this connection, a new series of tuftsin analogs containing D-amino acid residues in position *3* was investigated (28). It was found that D-Leu³-, D-Ala³-, and D-Arg³-tuftsin are inactive. Some biological activity however could not be excluded for D-Thr³ and D-Pro³-tuftsin.

From these data it is rather difficult to conclude where Pro³-residue interacts directly with the receptor by its γ-methylene group.

References

1. Najjar, V. A., 1974. Adv. Enzymology 41: 129–178.
2. Nishioka, K., Constantopoulos, A., Satoh, P. S., Mitchell, M. V. & Najjar, V. A., 1973. Biochim. Biophys. Acta 310: 217–229.

3. Nishioka, K., Satoh, P. S., Constantopoulos, A. & Najjar, V. A., 1973. Biochim. Biophys. Acta 310: 230–237.
4. Najjar, V. A., Chaudhuri, M. K., Konopińska, D., Beck, B. B., Layne, P. P. & Linehan, L., 1981. Augmenting Agents in Cancer Therapy (Hersh, E. M., Chirigos, M. A. & Mastrangelo, M., eds.), 459–478, Raven Press, New York.
5. Najjar, V. A., 1978. Exp. Cell Biol. 46: 114–126.
6. Najjar, V. A., 1970. Ger. Offen., 2, 123, 003, US Appl. 98, 860.
7. Najjar, V. A., 1980. Macrophages and Lymphocytes, (Escobar, M. R. & Friedman, M., eds.), part A, pp. 131–147, Plenum Publ. Corp.
8. Konopińska, D., Nawrocka, E., Siemion, I. Z., Szymaniec, S. & Slopek, S., 1976. Peptides 1976, Proc. 14th Europ. Pept. Symp., Wepion 1976. (Loffet, A. ed.), pp. 535–539, Ed. de l'Universite de Bruxelles, Bruxelles.
9. Konopińska, D., Nawrocka, E., Siemion, I. Z., Slopek, S., Szymaniec, S. & Klonowska, E., 1977. Int. J. Pept. Prot. Res. 9: 71–77.
10. Martinez J., Winternitz, T. & Vindel, J., 1977. Eur. J. med. Chem. - Chimica Therapeutica 12: 511–516.
11. Fridkin, M., Stabinsky, Y., Zakuth, V. & Spirer, Z., 1976. Peptides 1976, Proc. 14th Europ. Pept. Symp., Wepion 1976 (Loffet, A. ed.), pp. 541–550, Ed. de l'Université de Bruxelles, Bruxelles.
12. Fridkin, M., Stabinsky, Y., Zakuth, V. & Spirer, Z., 1977. Biochim. Biophys. Acta 496: 203–211.
13. Yajima, H., Ogawa, H., Watanabe, H., Fuji, N., Kurobe, M. & Miamoto, S., 1975. Chem. Pharm. Bull., 371–374.
14. Vičar, J., Gut, V., Frič, L. & Blaha, K., 1976. Coll. Czechoslov. Chem. Commun. 41: 3467–3473.
15. Najjar, V. A., 1979. Klin. Wochenschr. 57: 751–756.
16. Herman, Z. S., Plech, A., Stachura, S., Siemion, I. Z. & Nawrocka, E., 1981. Experientia, 37: 76–77.
17. Constantopoulos, A. & Najjar, V. A., 1973. J. Biol. Chem. 248: 3819–3822.
18. Spirer, Z., Zakuth, V., Bogair, N. & Fridkin, M., 1977. Eur. J. Immunol. 7: 69–74.
19. Koenig, W. & Seiler, F. R., 1975. Ger. Offen. 2, 343, 034; Chem. Abs. 83: 10892v.
20. Konopińska, D., Siemion, I. Z., Szymaniec, S. & Slopek, S., 1979. Polish J. Chem. 53: 343–351.
21. Stabinsky, Y., Gottlieb, P. D. & Fridkin, M., in press.
22. Edelman, G. M., Cunningham, B. A., Gall, W. E., Gottlieb, P. D. & Rutinshawaxdol, M. J., 1969. Proc. Natl. Acad. Sci. USA 63: 78–85.
23. Konopińska, D., 1978. Polish J. Chem. 52: 953–957.
24. Nair, R. M. G., Ponce, B. & Fudenberg, H. H., 1978. Immunochemistry 15: 901–907.
25. Yasumura, K., Okamoto, K. & Shimamura, S., 1977. Yakugaku Zasshi 97: 324–329.
26. Kopple, K. D., Go, A. & Pilipauskas, D. R., 1975. J. Am. Chem. Soc. 97: 6830–6838.
27. Zimmerman, S. S. & Scheraga, H. A., 1977. Biopolymers 16: 811–843.
28. Nawrocka, E., Siemion, I. Z., Slopek, S. & Szymaniec, S., 1980. Int. J. Pept. Prot. Res., 16: 200–207.
29. Venkatachalam, C. M., 1968. Biopolymers 6: 1425–1436.
30. Ovchinnikov, Yu. A., 1973. Peptides 1972, Proc. 12th Europ. Pept. Symp. (Hanson, H. & Jakubke, H. D. eds.) p. 10, North-Holland Publ. Co., Amsterdam.
31. Chou, P. Y. Fasman, G. D., 1977. J. Mol. Biol. 115: 135–175.
32. Hodes, Z. I., Nemethy G. & Scheraga, H. A., 1979. Biopolymers 18: 1611–1634.
33. Sucharda-Sobczyk, A., Siemion, I. Z. & Konopińska, D., 1979. Europ. J. Biochem. 96: 131–139.
34. Chandrasekaran, R., Lakshminarayan, A. V., Pandya, U. V. & Ramachandran, G. N., 1973. Biochim. Biophys. Acta 303: 14–27.
35. Sucharda-Sobczyk, A., Siemion, I. Z. & Nawrocka, E., 1980, Acta Biochim Pol 27: 353–363.
36. Siemion, I. Z., Lisowski, M., Konopińska, D. & Nawrocka, E., 1980. Eur. J. Biochem., 112: 339–343.
37. Grathwohl, Ch. & Wuthrich, K., 1976. Biopolymers 15: 2025–2041.
38. Deslauriers, R., Becker, J. M., Steinfeld, A. S. & Naider, F., 1979. Biopolymers 18: 523–538.
39. Blumenstein, M., Layne, P. P. & Najjar, V. A., 1979. Biochemistry 18: 5247–5253.
40. Sekacis, I. P., Lyepintch, E. E., Veretennikova, N. J. & Chipens, G. J., 1979. Bioorg. Kim. (russ.) 5: 1617–1622.
41. Bystrov, V. F., Ivanov, V. T., Portnova, S. L., Balashova, T. A. & Ovchinnikov, Yu. A., 1973. Tetrahedron 29: 873–876.
42. Nikoforovich, G. V., 1978. Bioorg. Khim. (russ.) 4: 1427–1430.
43. Konopińska, D., Nawrocka, E., Siemion, I. Z., Szymaniec, S. & Slopek, S., 1979. Arch. Immunol. Therap. Exper. 27: 151–157.
44. Hisatsune, K., Kobayashi, K., Nozaki, S. & Muramatsu, I., 1978. Chem. Pharm. Bull. 26: 1006–1007.
45. Stabinsky, Y., Fridkin, M., Zakuth, V. & Spirer, Y., 1979. Int. J. Pept. Prot. Res. 12: 130–138.
46. Fitzwater, S., Hodes, Z. I. & Scheraga, H. A., 1978. Macromolec. 11: 805–811.
47. Matsuura, S., Takasaki, A., Hiratani, H., Kotera, T. & Fujiwara, S., 1975. Jappan Kokai 75: 373; Chem. Abs. 83: 114937e.
48. Konopińska, D., Siemion, I. Z., Szymaniec, S. & Slopek, S., 1978. Polish J. Chem. 52: 573–580.
49. Konopińska, D., Siemion, I. Z. & Baros-Stuliglowa, E., 1978. Polish J. Chem. 52: 2255–2257.
50. Okamoto, K. & Shimamura, S., 1976. Yakugaku Zasshi 96: 315–320.
51. Yamanaka, N., Fukushima, M., Matsauka, H., Nishida, K. & Ota, K., 1979. Phagocytosis, pp. 93–100, Tokyo Press, Tokyo.
52. Kent, H. A., 1975. J. Biol. Reprod. 12: 504–509.
53. Chipens, G., Nikiforovich, G., Mutulis, F., Veretennikova, N., Vosekalna, L., Sosnov, A., Polevayz L., Ancans, J., Mishlakova, N., Liepinsh, E., Sekacis, I. & Breslav, M., 1979. Peptides, Structure and Biological Function, Proc. 6th American Pep. Symp. (Gross, E. & Meienhofer, Y., eds.), pp. 567–570, Pierce Chem. Comp.

Revision received January 15, 1981.

Macrophage activation by tuftsin and muramyl-dipeptide

M. Bruley-Rosset, I. Florentin and G. Mathé
Hôpital Paul-Brousse (Assistance Publique), Institut de Cancerologie et d'Immunogenetique (Inserm U50 & Association Claude-Bernard), 94800 Villejuif, France

Summary

Peritoneal macrophages from tuftsin or MDP-treated mice were tested for their cytostatic activity for tumor cell proliferation. Both substances are able to activate macrophages either after intravenous injection or after incubation in vitro with normal macrophages. But a stimulation as well as an inhibition of tumor cell growth can result from macrophage activation depending on the timing and dose injected. Restoration of the impaired cytostatic capacity of macrophages of mice observed with aging, is obtained by repeated administration of tuftsin. Normal and BCG-stimulated macrophages were examined for their regulatory activity on the proliferation of P815 tumor cells. Low density of macrophages per well determines a stimulation of target cell growth whether the macrophages are normal or activated. When the number of macrophages is increased, under conditions in which normal macrophages are not inhibitory. BCG-stimulated macrophages exert already a strong cytostatic activity. At high macrophage content it appears that normal macrophages can also display an inhibitory activity. Macrophage-tumor cell interactions are highly dependent on the concentration and the state of activation of macrophages.

Introduction

Among immunological functions susceptible to stimulation by active peptides, macrophage activation is an important effector mechanism to be considered since there is much evidence of its role in resistance to malignant tumors (1) and to infections caused by intracellular pathogens (2). Macrophage activation has a specific immunological basis but its expression is nonspecific (3). The process of activation is a multiple step phenomenon. At different degrees of stimulation depending on the nature of the stimulus, correspond various morphological, biochemical and functional changes (4–6). A property of activated macrophages is their capacity for recognizing and destroying neoplastic cells (7). The use of acquired tumoricidal function has been considered in this work to indicate that macrophage activation was present after administration of two peptides: tuftsin (8, 9) and muramyl-dipeptide (MDP) (10, 11).

In this presentation we have examined the capacity of macrophages from treated animals to exert a cytostatic activity for tumor cells in relation to the dose and the time of peptide administration. We also looked at the capacity of tuftsin given repeatedly to restore the impaired function of macrophages observed in aged mice.

In a second part, we added a study on the role of macrophages activated by BCG on the regulation of tumor cell proliferation in relation to the state of activation and to culture conditions.

Material and methods

Mice

3-month-old C57BL/6 or (C57BL/6 × DBA/2)

Molecular and Cellular Biochemistry 41, 113–118 (1981).   0300-8177/81/0041-0113/$01.20.

114

F1 female mice were purchased from OLAC Laboratories (England), 12-month-old C57BL/6 female mice, used in the restorative studies, were purchased from Bom Holtgard Laboratories (Denmark).

Peptides

Muramyl-dipeptide = MDP: Ac-Mur-L-Ala-D-Glu-NH$_2$ prepared by Lederer (10) was given at the dose of 500 μg or 100 μg per mouse.

Tuftsin: L-Thr-L-Lys-L-Pro-L-Arg prepared by Martinez (from Winternitz Laboratory, France) was given at the dose of 25 or 10 μg per mouse.

A single injection of the different doses was given intravenously 7, 3, or 1 days before testing. For the restorative studies, 12-month-old mice were injected (i.p.) weekly with 10 μg of tuftsin during a period of six months.

Assay for macrophage-mediated cytostatis

Peritoneal macrophages were harvested after peptide administration and distributed into microplates at the concentration of 2 × 10⁶ cell/ml in RPMI 1640 medium. After 1 h of incubation at 37 °C, non-adherent cells were removed by vigorous washing. Tumor cells (P815) were then added to macrophage monolayers at the optimal ratio of 1 target cell for 20 macrophages. After 18 h of incubation; 1 μCi/well of ³H-TdR was added for the last 4 h of culture, and thymidine incorporated into tumor cells was measured in a β counter (Protocol 1).

For some experiments, in order to avoid competition with cold thymidine released by activated macrophages, tumor cells, after 18 h of incubation in the presence of macrophages, were harvested, washed, diluted in fresh medium, and redistributed in a new microplate (Protocol 2). Thymidine was added as in Protocol 1. Results are expressed as mean cpm of ³H-TdR incorporation ± standard error of six cultures and as percent inhibition (–) or stimulation (+) of tumor cell proliferation.

Macrophage activation in vitro

Peritoneal macrophages from normal mice were cultivated on monolayers with 1 or 0.1 μg of tuftsin or 10 or 1 μg of MDP. After 48 h of culture, tumor cells were added and the cytostatic test was performed as previously described.

Results

1. Effect of in vivo administration of tuftsin and MDP on macrophage cytostatic activity of young mice

Peritoneal macrophages are collected 3, 7 and 10 days after i.v. administration of tuftsin at the dose of 25 μg or MPD at the dose of 500 μg.

Macrophages activated by tuftsin exert a marked cytostatic effect already at day 3 (40% inhibition of tumor cell growth) and this activity is maintained until the 10th day after the injection (Table 1).

A slight activation of macrophages by MDP is observed 7 days after its administration (26% inhibition). An opposite effect, i.e. an increased tumor cell growth is obtained at the other times (day 3 and day 10; Table 2).

Table 1. Cytostatic activity of peritoneal macrophages from young mice at different times after administration of 25 μg of tuftsin.

	Untreated mice	Day of i.v. administration of 25 μg of tuftsin		
		Day 3	Day 7	Day 10
Mean number of cpm of ³H-TdR ± SE incorporated into tumor cells in the presence of macrophages	94 812 ± 5 148	58 446 ± 4 134	50 338 ± 3 271	55 079 ± 3 851
% inhibition (–) of incorporation compared to controls		(–) 38%	(–) 47%	(–) 42%

Table 2. Cytostatic activity of peritoneal macrophages from young mice at different times after administration of 500 μg of muramyl-dipeptide.

Mean number of cpm of ³H-TdR ± SE incorporated into tumor cells in the presence of macrophages from:	Day of administration of 500 μ of MDP		
	Day 3	Day 7	Day 10
Untreated mice	8 941 ± 867	12 632 ± 1 102	8 054 ± 699
MDP-treated mice	22 732 ± 692	9 386 ± 890	10 400 ± 1 347
% inhibition (-) or stimulation (+) of incorporation compared to controls	(+) 254%	(-) 26%	(+) 129%

Table 3. In vitro activation of normal peritoneal macrophages after 48 h of culture with tuftsin or muramyl-dipeptide.

		Normal macrophages incubated 48 h in the presence of			
		Tuftsin		MDP	
		0.1 μg	1 μg	1 μg	10 μg
Mean number of cpm of ³H-TdR ± SE incorporated into tumor cells in presence of macrophages	7 208 ± 1 769	2 488 ± 535	1 671 ± 711	2 350 ± 218	8 771 ± 905
% inhibition (-) or stimulation (+) of tumor cell proliferation compared to controls	-	(-) 66%	(-) 77%	(-) 68%	(+) 121%

2. Effect of in vitro incubation of tuftsin and MPD on macrophage cytostatic activity

Tuftsin and MDP are added at different doses to a culture of normal macrophages; after an in vitro incubation of 48 h the cytostatic test is performed. Macrophages are strongly activated by tuftsin: 66% and 77% inhibition at doses of 0.1 μ and 1 μg respectively (Table 3). MDP is able to induce a cytostatic activity of normal macrophages only after incubation with 1 μg. The high dose (10 μg) leads to a stimulation of tumor cell proliferation rather than an inhibition (Table 3).

3. Effect of repeated administrations of tuftsin on macrophage cytostatic activity of aged mice

This experiment is performed according to two protocols: Protocol 2 is performed in addition to the usual Protocol 1, in order to determine the influence of the cold thymidine released by activated macrophages on the incorporation of tritiated thymidine by tumor cells (Table 4). The first observation is the decreased capacity of aged macrophages to control tumor cell proliferation compared to young macrophages in both experimental conditions. The weekly administration of tuftsin (10 μg) to C57BL/6 mice during a period of 6 months (until the age of 18 months) is able to restore this impaired cytostatic capacity of macrophages. The same results are observed by using the two different protocols, but the degree of inhibition is lower in Protocol 2 demonstrating competition between cold and tritiated thymidine incorporation into tumor cells.

4. Influence of the density and the degree of macrophage activation on tumor cell proliferation in vitro

In this experiment, macrophages are activated by an i.v. injection of 1 mg of BCG 14 days before

Table 4. Cytostatic activity of peritoneal macrophages from aged mice repeatedly treated with tuftsin (10 μg).

Mean number of cpm of ³H-TdR ± SE incorporated into tumor cells	Macrophages from young mice	Macrophages from untreated aged mice	Macrophages from tuftsin (10 μg) treated aged mice	Without macrophage
Protocol 1	2 166 ± 262	14 506 ± 1 124	665 ± 77	67 778 ± 7 176
% inhibition (–) or stimulation (+) compared to young untreated controls	0%	(+) 670%	(–) 70%	–
% inhibition compared to aged untreated controls	(–) 85%	0%	(–) 95%	–
Protocol 2	8 660 ± 698	28 664 ± 1 061	10 698 ± 810	30 300 ± 2 585
% inhibition (–) or stimulation (+) compared to young untreated controls	0%	(+) 330%	(+) 20%	–
% inhibition compared to aged untreated controls	(–) 70%	0%	(–) 63%	–

Protocol 1: the incorporation of ³H-TdR by tumor cells is performed in the presence of macrophages.
Protocol 2: the incorporation of ³H-TdR by tumor cells is performed in the absence of macrophages after the transfert of tumor cells into a new microplate.

testing. The tumor cell proliferation is assayed in the presence of different numbers of normal or BCG-activated macrophages, the number of tumor cells being constant (5 × 10⁴/well). We have first observed a stimulation of tumor cell growth,

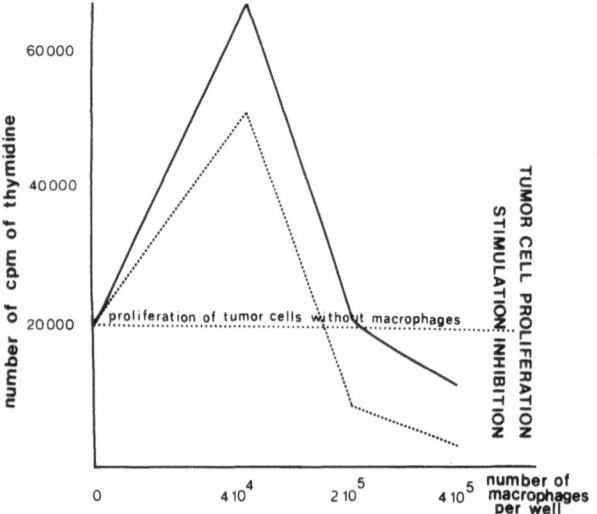

Fig. 1. Proliferation of tumor cells (P815) measured by tritiated thymidine incorporation in the absence (.....) or in the presence of normal (——) or BCG-activated (----) macrophages in relation to their concentration.

measured by thymidine incorporation, when the number of macrophages per well is low (4 × 10⁴) in the presence of normal macrophages as well as activated macrophages, but in the latter case, the degree of stimulation is lower. At a macrophage concentration of 2 × 10⁵/well, thymidine incorporation is reduced to about 50% by BCG-activated macrophages whereas, at the same concentration normal macrophages do not exert any inhibitory activity. The degree of inhibition observed with BCG-activated macrophages (75%) increases with the number of macrophages per well (4 × 10⁵) and even normal macrophages can reduce tumor cell proliferation but at a lesser extend (40%) (Fig. 1).

Discussion

The capacity of two peptides, tuftsin and MDP, to induce peritoneal macrophage activation has been examined after *in vivo* i.v. administration or after *in vitro* incubation with normal macrophages. A number of observations have emerged from this work:

(1) The degree of activation depends on the dose

and timing of peptide injections as it has been demonstrated for several other adjuvants (12), Tuftsin is known for its capacity to stimulate phagocytosis (13) and to render macrophages cytotoxic (14). Other investigators have demonstrated that MDP significantly enhanced both phagocytosis and killing of intracellular bacteria (15). Macrophage cytotoxicity could be obtained after *in vitro* and *in vivo* stimulation (16).

(2) The fact that tuftsin and MDP are able to stimulate normal macrophages to become cytostatic in vitro suggests that in this case the participation of host T lymphocytes is not necessary in the activation process. Such a phenomenon has already been described for activation by double-stranded RNA and LPS (16). Furthermore, macrophage activation can be obtained in congenitally athymic nude mice (17).

(3) Tuftsin is not only able to induce an increased tumoricidal activity of macrophages from young mice but is also able to restore the cytostatic capacity of macrophages which is impaired in mice with advancing age. This experiment points out that the release of cold thymidine by activated macrophages may contribute to the inhibition of tumor cell growth but is certainly not the unique factor responsible for growth inhibition.

(4) Inhibition or stimulation of target cell proliferation can be obtained, and these opposite effects are dependent on the experimental conditions, timing of MDP administration *in vivo,* or dose of MDP added *in vitro*. We have observed that low doses and short time intervals after *in vivo* administration of BCG lead to enhancement of proliferation (personal observations). This stimulation of tumor cell multiplication always preceeds or follows an inhibition and seems to reflect a weak level of macrophage activation.

To understand better this dual phenomenon, normal and BCG-stimulated macrophages were tested for their regulatory activity on tumor cell proliferation according to the culture conditions. We observed a biphasic effect of macrophages according to their degree of activation and their density in the culture plates. In wells containing sparse macrophage monolayers both normal and BCG-stimulated macrophages stimulate P815 target growth. But under conditions in which normal macrophages are not inhibitory, BCG-stimulated macrophages exert already a strong

cytostatic activity. At a relatively high macrophage content, it appears that normal macrophages can also display an inhibitory effect on tumor cell proliferation, but a much higher number of normal than of activated macrophages is required to obtain a similar effect.

Contrasting with the number of papers showing that activated macrophages appear to kill or inhibit the growth of malignant cells, the data given by Nathan (18) suggest that BCG-activated macrophages are only much less stimulatory than normal macrophages. Identic results to our findings were found by using thioglycollate, Con A or BCG-stimulated macrophages or normal macrophages which either enhance or inhibit target cell proliferation depending on the density of macrophages (19). The experiments performed by Keller (20) attest to the inherent capacity of macrophages to block the proliferation of any rapidly dividing cells.

All these observations suggest that the macrophage-tumor cell interaction may depend on the ratio of effectors and targets. Macrophages are also known to modulate the responses of lymphocytes to variety of stimuli. Macrophages appear to play an important role in host homeostatic regulation of cell proliferation by its state of activation and its concentration at the site of the reaction.

Acknowledgments

We wish to thank Dr. Chedid for the supply of muramyl-dipeptide and Dr. Martinez for the preparation and the gift of tuftsin.

References

1. Schultz, R. M., Papamatheakis, D., Luetzeler, J. & Chirigos, M. A., 1977. Cancer Res. 37: 3338–3347.
2. Mackaness, G. B., 1962. J. Exp. Med. 116: 381–388.
3. Mackaness, G. B., 1969. J. Exp. Med. 129: 973–985.
4. North, R. J., 1978. J. of Immunol. 121: 806–809.
5. Karnosky, M. L. & Lazdins, J. K., 1978. J. of Immunol. 121: 809–813.
6. Cohn, Z. A., 1978. J. of Immunol. 121: 813–816.
7. Hibbs, J. B., Lambert, L. H. & Remington, J. S., 1972. Proc. Soc. Exp. Biol. Med. 139: 1049–1052.
8. Martinez, J., Winternitz, F. & Vindel, J., 1980. Eur. J. Med. Chem., in the press.
9. Florentin, I., Bruley-Rosset, M., Kiger, N., Imbach, J. L., Winternitz, F. & Mathé, G., 1978. Cancer Immunol. Immunoth. 5: 211–216.

118

10. Ellouz, F., Adam, A., Ciobaru, R. & Lederer, E., 1974. Biochem. Biophys. Res. Commun. 59: 1317–1325.
11. Juy, D. & Chedid, L., 1975. Proc. Natl. Acad. Sci. USA 72: 4105–4111.
12. Bruley-Rosset, M., Florentin, I., Khalil, A. M. & Mathé, G., 1976. Int. Arch. Allergy Appl. Immun. 51: 594–607.
13. Najjar, V. A. & Nishioka, K., 1970. Nature (London) 228–673.
14. Nishioka, K., 1979. Brit. J. Cancer 39: 342–345.
15. Hadden, J. W., Englard, A., Sadlik, J. R. & Hadden, E. M., 1979. Int. J. Immunopharmacol. 1: 17–27.
16. Alexander, P. & Evans, R., 1971. Nature New Biol. 232: 76–78.
17. Cheers, C. & Waller, R., 1975. J. of Immunol. 115: 844–847.
18. Nathan, C. F. & Terry, W. D., 1975. J. Exp. Med. 142: 887–902.
19. Goldman, R. & Bar-Shavit, Z., 1979. J. Natl. Cancer Inst. 63: 1009–1016.
20. Keller, R., 1974. Brit. J. Cancer 30: 401–415.

Received August 11, 1980.

Reprint requests should be addressed to M. Bruley-Rosset

Induction of T lymphocyte differentiation by thymic factors

Jean-Louis Touraine

Transplantation and Immunobiology Unit, INSERM U 80 and CNRS ERA 782, Pav. P., Hôpital E. Herriot, 69374 Lyon Cedex 2, France

Introduction

Several, biochemically distinct, factors of thymic origin have been isolated and characterized (1–4). These factors have been shown to be active in some immunological assays in vitro and in vivo. In particular they can induce the appearance of surface characteristics of T cells onto murine or human prothymocytes (5, 6). These thymic inducers probably account for the partial immune reconstitution of neonatally thymectomized mice following a gestation (7) or the implantation of cell-impermeable Millipore diffusion chambers containing thymic tissue (8).

Several in vitro and in vivo properties of T lymphocytes can be induced or enhanced by previous incubation with thymic factors. None of the presently available preparations, however, is capable of fully reconstituting the immunity of a neonatally thymectomized mouse or a patient with Di George syndrome. This observation contrasts with the total, dramatic and persistent reconstitution obtained with fetal thymus transplants. Besides the potent differentiating effect of thymic factors it may thus be postulated that there exists additional mechanisms for the induction of a T-cell developmental program (9).

Materials and methods

Thymic factors

Calf thymuses were provided by a local slaughter house and pieces of human thymuses were obtained from young children undergoing cardiac surgery. Extracts were prepared according to the procedure of A. L. Goldstein et al. (1). The human and calf thymosins-fraction 5 thus obtained were used in vitro at the protein concentration of 100–125 μg/ml. Human spleen extracts were similarly prepared and used as a control. FTS was a gift from J. F. Bach and M. Dardenne. It was used at the concentration of 1 ng/ml.

Thymic epithelial culture supernatants

Pieces of human thymuses from children or fetuses (approximately 8–10 weeks old) were cut into small fragments and cultured on a steel mesh in plastic Petri dishes containing RPMI-1640 medium, supplemented with 2 mM glutamine and 15% inactivated AB human serum. The medium was changed every three days. The cultures were maintained, at 37 °C and in a humidified atmosphere of 5% CO_2 – 95% air, until virtually no thymocytes were found in the supernatant. The medium was then removed and replaced by serum-free medium. After 24 h, the serum-free supernatant was collected and stored at -25 °C until use in the various assays. The thymic cells were then examined by light and electron microscopy. Because thymuses from young fetuses have few lymphoid cells, cultures were shorter in these cases than when thymuses of children were used. As controls, supernatants of cultured spleen fragments and kidney fragments were prepared.

Cell suspensions

Human peripheral blood lymphocytes were

Molecular and Cellular Biochemistry 41, 119–122 (1981).   0300-8177/81/0041–0119/$00.80.

© 1981, Martinus Nijhoff/Dr W. Junk Publishers, The Hague.

120

separated from heparinized venous blood by centrifugation on a Ficoll-Hypaque gradient (10). Human bone marrow cells were separated in five layers by centrifugation on a discontinuous density gradient of bovine serum albumin (BSA) as described (11). Thymocytes were prepared from thymuses of children undergoing cardiac surgery.

Induction assays

The induction of the HTLA$^+$ phenotype (human T lymphocyte differentiation antigens) was performed by incubating equal volumes of marrow cell suspension (2 × 10^6 cells/ml) from each layer with thymic factors, control factors or supernatants, for 2 h at 37 °C. Cells were washed and the percentage of HTLA$^+$ cells was routinely determined using a microcytotoxicity method with a specific anti-HTLA serum and rabbit complement (6, 11). Occasionally, HTLA were also shown at the cell surface using an immunofluorescence method or immunoperoxydase labeling. OKT 4$^+$ cells were identified by immunofluorescence. The induction of the capability to form E-rosettes was performed over a 6 h incubation with the various factors. The enhancement of the response to mitogens or to allogeneic stimuli involved either a pre-incubation of cells with the factors for 18 h or a simultaneous addition and coincubation of the stimulant and the factors to the cell culture.

Lymphocyte cultures

The cultures of lymphocytes with phytomitogens or mitomycin-treated allogeneic cells were performed using a micromethod derived from a previous technique (12). Proliferation was evaluated by the amount of 3H-thymidine incorporated during the last 24 h of the culture. For evaluation of suppressor T-cell activity, T-lymphocytes were cultured for three days with concanavalin A (Con A), the lymphoblasts were then separated on

Table 1. Some immunologically active thymic factors.

Factors	Authors
Thymosins α_1, α_7, β_3, β_4, β_1	A. L. Goldstein
Thymopoietin I, II	G. Goldstein
Facteur thymique sérique (FTS)	J. F. Bach
Thymic humoral factor (THF)	N. Trainin

Table 2. Conversion of human bone marrow prothymocytes into HTLA$^+$ lymphocytes upon incubation with thymic factors.

Factors	% HTLA$^+$ Cells in bone marrow layer				
	I	II	III	IV	V
Control RPMI-medium	3	5	5	7	6
Human splenic factors	5	8	10	8	6
Human thymic factors	5	14	26	10	5
Human thymic supernatant	6	17	25	8	7

Bone marrow cells were incubated for 2 h at 37 °C with human thymic factors (prepared similarly as thymosin-fraction 5), with serum-free supernatant of thymic epithelial cells in culture, or with control solutions. A significant conversion of cells to the HTLA$^+$ phenotype regularly occurred in layers II and III, after incubation with either thymic extracts or thymic supernatants (means of results obtained with bone marrow cells from ten normal marrow donors).

a BSA gradient and added to a mixed lymphocyte culture (MLC). Various concentrations of untreated or mitomycin-treated lymphoblasts were introduced into the MLC and the degree of suppression of the mixed lymphocyte reaction was calculated as percent inhibition of control lymphocyte proliferation.

Results and discussion

The various thymic factors

Table 1 mentions some of the thymic factors which have been described and shown to be active in some assays. The multiplicity of the thymic peptides, whether isolated from the thymus itself or from blood, is a matter for some confusion. The requirements for a thymic hormone have been recently defined (13): purity, structure and synthesis, activity in relevant assays, specificity of site of production by the thymus, demonstration of receptor molecules on appropriate target cells and T-cell selectivity. Most of the above-mentioned thymic factors fulfill these criteria. Yet they have completely different chemistries. Additional factors of thymic origin, including those secreted by thymic epithelial cells in culture supernatants (9, 14), have comparable activities. The significance of this plurality of thymic factors is still unknown. It is, however, obvious that they do not simply derive one from the other and that they are not merely active one after the other to achieve the stepwise differentiation of T lymphocytes.

Induction of T-cell surface markers

Thymic factors can induce the appearance of T-cell characteristics onto prothymocytes from the bone marrow or the spleen (5, 6). For instance some mouse cells are induced to express Thy-1$^+$ phenotype or to form azathioprine-sensitive rosettes following incubation with thymosin, thymopoietin or FTS. Human prothymocytes present in layers II and III of BSA-separated bone marrow cells are converted into cells expressing the HTLA$^+$ phenotype, as determined by cytotoxicity or immunofluorescence, after 2 h of incubation with thymic factors (Table 2). This is an active phenomenon requiring active metabolic conditions.

Induction of the ability to form E-rosettes also occurs but one should carefully distinguish the minor increase observed immediately after addition of extracts of various tissues (probably due to improved conditions for the binding of sheep erythrocytes) from the more significant augmentation noticed 6 h or more later (when a more profound modification of the cells of T-lineage has occurred).

A variety of other T-cell differentiation markers has been shown to be inducible by thymic factors, including the TL and Lyt 1, 2 and 3 alloantigens in the mouse, as well as the terminal deoxynucleotidyl transferase (TdT).

In human the OKT antigens may also be inducible, as suggested by our observation in the peripheral blood of a patient with a relatively selective deficiency in OKT 4$^+$ lymphocytes. After incubation with FTS for 2 h, 11.5% of peripheral blood lymphocytes were converted from the OKT 4$^-$ to the OKT 4$^+$ phenotype, as demonstrated in immunofluorescence (Table 3). Whether, in this case, the modification has mainly been a surface rearrangement or a more profound change in the cell is still uncertain.

All these changes occur only in cells committed to the T-linearge. They are not equivalent to the initial induction of a T-cell developmental program in an uncommitted stem cell. Several other agents especially those which increase intra-cellular levels of cyclic AMP, also induce the appearance of some T-cell characteristics at the surface of prothymocytes (15).

Induction or enhancement of in vitro properties

An increased proliferative response to mitogens and to allogeneic cells has been observed following pretreatment of bone marrow cells, thymocytes (16) or, to a lesser degree, peripheral blood lymphocytes (9) with thymic factors. Supernatants of thymic epithelial cultures also enhanced significantly these proliferative responses when used for a pretreatment of thymocytes or when added simultaneously with the stimulant. Supernatants from fetal thymuses were the most effective and the control factors or supernatants induced either little or no alterations of the proliferative responses. In the above-described suppressor assay, 5 to 25×10^3 Con A lymphoblasts could decrease the allogeneic response (4×10^7 lymphocytes) by more than 50% (17). The effects of the thymic factors and supernatants varied from an increased to a completely abrogated suppressor activity depending on the dose, the timing, and the cell origin. Increased (18) and decreased (19) suppressor activities have been previously observed in different experimental conditions, using thymosin preparations or supernatants. Enhancement of helper T-cell function or of T-cell mediated cytotoxicity after in vitro incubation with thymic factors has been less frequently reported.

From these and other (12) results, as well as ontogenetic studies, we schematically envisioned human T-cell differentiation as involving several sequential stages, including: HTLA, then capability to form E-rosettes, and later bifurcational development resulting in allogeneic-responsive cells and Con A-PHA-responsive cells (16, 17). More complexity is introduced into the scheme by the numerous interactions between T-cell subsets and by the development of T-lymphocytes with dif-

Table 3. Enhancement of OKT 4$^+$ lymphocytes in peripheral blood of an OKT 4$^+$ deficient patient by incubation with FTS.

Factor	% of OKT 4$^+$ PBL
Control RPMI-medium	15.0
Inactive tissue extract	13.0
FTS	26.5

The patient's PBL were incubated with the appropriate factor for 2 h at 37 °C. OKT 4$^+$ cells were then enumerated, using an immuno-fluorescence method. The percentage of OKT 4$^+$ lymphocytes was found higher after incubation with FTS than in the other tubes, although it did not reached completely normal values.

ferent recognition structures, repertoires, and functions (20).

Increased T-cell functions in vivo

Adult-thymectomized animals progressively develop a partial T-cell deficiency. Treatment of these animals with thymic factors has resulted in restoration of normal T-cell numbers and improvement of many T-cell functions.

In various models, increased helper, suppressor and cytotoxic T cell functions after in vivo treatment with thymic factors have been reported (21, 22).

A full reconstitution of a completely T-cell deficient animal or human has, however, not yet been achieved with cell-free thymic factors (9).

Although less efficient than thymic transplants, injections of thymic factors into patients has already shown very promising results, especially in partial T-cell deficiencies, be they primary or secondary diseases (23). As could be expected, the patients with the best results are those with a low serum level of thymic factor activity and with presence of lymphocytes converted into HTLA[+] cells after in vitro incubation with thymic inducers. The efficacy of thymic peptides requires the presence of cells of the T lineage susceptible to differentiate and/or proliferate under their influence. All these results may suggest that initial development of the T-cell lineage in vivo requires interactions between lymphoid cells and thymic reticulo-epithelial cells. In these interactions, the role of major histocompatibility complex determinants can be envisioned. The presently known thymic peptides would be active after this initial 'programmation' both in the thymus, at high concentration, and in the periphery with a distinct mechanism of action.

Acknowledgments

F. Touraine's contribution to several of the reported studies is gratefully acknowledged. I also thank O. de Bouteiller for excellent technical assistance. These investigations were supported by grants from the 'Délégation Générale à la Recherche Scientifique et Technique' and the 'Centre National de la Recherche Scientifique'.

References

1. Low, T. L. K., Thurman, G. B., Chincarini, C., McClure, J. E., Marshall, G. D., Hu, S. K. & Goldstein, A. L., 1979. Ann. N.Y. Acad. Sci. 332: 33–48.
2. Schlesinger, D. H., Goldstein, G. & Niall, H. D., 1975. Biochemistry 14: 2214–2218.
3. Bach, J. F., Dardenne, M., Pleau, J. M. & Rosa, J., 1977. Nature (London) 266: 55–57.
4. Trainin, N., Umiel, T. & Yakir, Y., 1980. In: Thymus, Thymic Hormones and T lymphocytes, Ed. F. Aiuti & H. Wigzell, Academic Press, London, pp. 201–211.
5. Komuro, K., Boyse, E. A., 1973. Lancet, 1: 740–743.
6. Touraine, J. L., Incefy, G. S., Touraine, F., Rho, Y. M. & Good, R. A., 1974. Clin. Exp. Immunol. 17: 151–158.
7. Osoba, D., 1965. Science 147: 298–299.
8. Osoba, D. & Miller, J. F. A. P., 1964. J. Exp. Med. 119: 177–194.
9. Touraine, J. L., Touraine, F., 1979. Ann. N.Y. Acad. Sci. 332: 64–69.
10. Boyum, A., 1968. Scand. J. Clin. Lab. Invest. 21: (suppl. 97) 9–109.
11. Touraine, J. L., Touraine, F., Incefy, G. S. & Good, R. A., 1975. Ann. N.Y. Acad. Sci. 249: 335–342.
12. Touraine, J. L., Touraine, F., Hadden, J. W., Hadden, E. M. & Good, R. A., 1976. Int. Arch. Allergy Appl. Immunol. 52: 105–117.
13. Bach, J. F. & Goldstein, G., 1980. Thymus 2: 1–4.
14. Kruisbeek, A. M., 1979. Ann. N.Y. Acad. Sci. 332: 109–112.
15. Sheid, M. P., Hoffman, M. K., Komuro, K., Hammerling, U., Abbott, J., Boyse, E. A., Cohen, G. H., Hooper, J. A., Schulof, R. S. & Goldstein, A. L., 1973. J. Exp. Med. 138: 1027–1032.
16. Touraine, J. L., Hadden, J. W. & Good, R. A., 1977. Proc. Nat. Acad. Sci. USA 74: 3414–3418.
17. Touraine, J. L., 1978. In: Human Lymphocyte Differentiation: Its Application to Cancer. B. Serrou & C. Rosenfeld, eds. INSERM Symposium 8, Elsevier Press North Holland, Amsterdam, pp. 93–100.
18. Horowitz, S., Borcherding, W., Moorthy, A. V., Chesney, R., Schulte-Wisserman, H. & Hong, R., 1977. Science, 197: 999–1004.
19. Serrou, B., Rosenfeld, C., Caraux, J., Thierry, C., Cupissol, D. & Goldstein, A. L., 1979. Ann. N.Y. Acad. Sci. 332: 95–100.
20. Touraine, J. L., 1980. In: Thymus, Thymic Hormones and T lymphocytes. F. Aiuti and H. Wigzell, eds. Academic Press, London, pp. 365–373.
21. Bach, J. F., Bach, M. A., Charreire, J., Dardenne, M., Pleau, J. M., 1979. Ann. N.Y. Acad. Sci. 332: 23–32.
22. Barker, A. D., Dennis, A. J., Moore, V. S. & Rice, J. M., 1979. Ann. N.Y. Acad. Sci. 332: 70–80.
23. Touraine, J. L. & Good, R. A., 1981. Thymus (in press).

Bactericidal activity of tuftsin

Jean Martinez and François Winternitz
Equipe de Recherche No. 195 CNRS, Ecole Nationale Supérieure de Chimie, 8, rue de l'Ecole Normale, 34075 Montpellier Cedex, France

Summary

The biological activities of the phagocytosis stimulating tetrapeptide, Thr-Lys-Pro-Arg are discussed. A brief account on the stimulation by tuftsin of phagocytosis of various particles, including bacteria was reported. Stimulation of bactericidal activity by this tetrapeptide was investigated *in vitro* as well as *in vivo*. The potency of tuftsin to enhance blood clearing of *Staphylococcus aureus*, *Listeria monocytogenes*, *Escherichia coli* and *Serratia marcescens* by mouse peritoneal macrophages was demonstrated.

Bactericidal activity and effects of tuftsin on this phenomenon were studied in liver and spleen of mice. Tuftsin stimulates these activities. Same experiments were performed in infected leukemic mice by *Serratia marcescens* or *Escherichia coli*. Results on blood clearing and bactericidal activities in liver and spleen were reported and compared to those of healthy and leukemic untreated animals. Tuftsin was found to present interesting stimulatory effects on the bactericidal activity of phagocytes.

Introduction

There are many defense mechanisms available to the host in its struggle with invading micro-organisms. One of this defense mechanisms, phagocytosis, plays an important role in the critical early stages of bacterial infections and, thereby, is a significant determinant of the eventual outcome of these infections. Effective phagocytic activity early in the course of bacterial invasion may limit the spread of bacteria and prevent ongoing infection, while ineffective phagocytosis may lead to uncontrolled bacterial multiplication and overwhelming infection (1).

A number of different cells types possess phagocytic capacity. Some, such as the circulating polymorphonuclear leucocytes (PMN) and monocytes, are able to migrate to the site of bacterial lodgement.

In a series of communications by NAJJAR and co-workers (2–4) it was demonstrated that a cytophilic-γ-glubulin fraction, leukokinin, of either human or canine origin, binds specifically and reversibly to autologous polymorphonuclear leucocytes and stimulates their phagocytic activity against pathogenic *Staphylococcus aureus*.

Soon thereafter, it was found that the whole stimulatory effect of leukokinin can be ascribed to a single tetrapeptide fragment (5, 6), christened tuftsin (6), whose sequence had been determined as Thr-Lys-Pro-Arg (7). Two distinct enzymes are engaged in the release of tuftsin from leukokinin (8, 9). One, leukokininase, a specific enzyme on the outer membrane of the phagocyte, cleaves the amino terminal end of the tetrapeptide. The other, which has not been yet properly characterized seems to be located in the spleen and cleaves at the carboxy end.

Tuftsin has been synthesized in various laboratories and its biological action, stimulation of phagocytosis of PMN and macrophages ascertained.

The potential eventuality of using tuftsin as a

Molecular and Cellular Biochemistry 41, 123–136 (1981).   0300–8177/81/0047–0123/$02.80.

124

drug against various infectious diseases prompted us to study some aspects of its stimulation of bactericidal activity. Because of direct correlation between increased phagocytic activity and enhancement of the cell's bactericidal activity (10, 11) and because of reduction of nitro blue tetrazolium (NBT), is a well known method of studying of at least one important aspect of this bactericidal activity (12–14), we will herein give some accounts on the effects of tuftsin upon these processes.

In vitro studies

Phagocytosis stimulation

Materials and methods

Assays are usually performed with human or dog PMN leucocytes, with PMN cells from guinea pig peritoneal exudates, as well as with macrophages from mouse peritoneal and rabbit lung washings.

These assays may involve phagocytosis of opsonized bacterial particles (*Staphylococcus aureus, Listeria monocytogenes*) (7–15, 18), heat-killed yeast particles (*Saccharomyces oviformis*) (20–21) or even latex particles.

The other type of phagocytosis assay involves the engulfment of particulate nitro blue tetrazolium and its further reduction through the glucose-6-phosphate pathway. The reduction of the yellow NBT to the corresponding formazan is one of the assays used to evaluate this stimulation (11). The reduced dye is measured at 515 nm (19, 20, 22).

For details, refer to original articles.

Results and discussion

As can be seen in Table 1, tuftsin highly stimulates phagocytosis of various particles including bacteria by PMN and macrophages. Biological activity of tuftsin has been demonstrated to be concentration dependent. Half maximum activation was obtained

Table 1. Stimulation of Phagocytosis by tuftsin

Phagocyte	Source of phagocyte	Particle	Tuftsin µg/ml	Index Δ	Ref.
Peritoneal macrophages	mouse	*Staphylococcus aureus*	0.1	1.8	4
Lung macrophages	rabbit	*Staphylococcus aureus*	0.1	1.9	4
Peritoneal PMN	guinea pig	*Staphylococcus aureus*	0.1	1.7	4
PMN leucocytes	dog	*Staphylococcus aureus*	0.1	1.8	4
PMN leucocytes	human blood	*Staphylococcus aureus*	0.05	1.9	18
Peritoneal PMN	guinea pig	*Staphylococcus aureus* 209 p	0.15	1.7	31
Peritoneal macrophages	mouse	*Staphylococcus aureus* S 57	0.01	1.3	11
Peritoneal macrophages	mouse	*Staphylococcus aureus* Londres	0.01	1.2	11
Peritoneal macrophages	mouse	*Staphylococcus aureus* Smith	0.01	1.3	11
Peritoneal macrophages	mouse	*Listeria monocytogenes*	0.01	1.1	11
Peritoneal PMN	human blood	*Saccharomyces oviformis*	0.1	1.7	9

Index $\Delta = \dfrac{\% \text{ of phagocyte cells containing particles with tuftsin}}{\% \text{ of phagocytes cells containing particles without tuftsin}}$

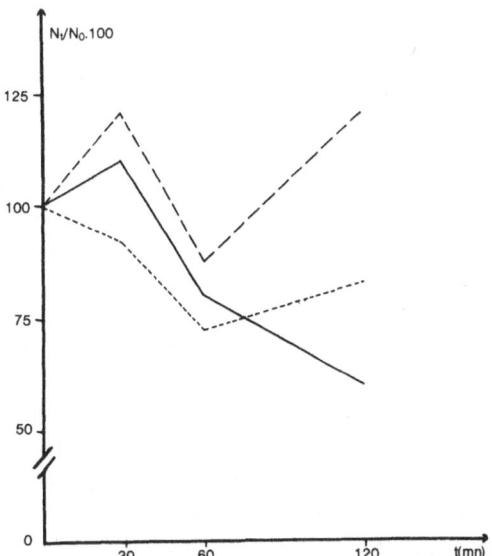

Fig. 1. Phagocytosis stimulation by tuftsin of *Listeria mono-cytogenes* by mouse PMN (Van Furth's technic). (— — control. —— normal serum. - - - tuftsin treated mice).

Table 3. Effect of [Des-Thr1]-tuftsin on the phagocytosis by mouse peritoneal macrophages*
[Des-Thr1]-tuftsin = 5 μg/ml

	P_1	P_2	Index Δ
Staphylococcus aureus Londres	28.3	30.1	0.94
Staphylococcus aureus Smith	8.9	9.9	0.90
Lysteria monocytogenes	37.1	47.1	0.79

Index $\Delta = \dfrac{P_1}{P_2} =$

$\dfrac{\% \text{ of phagocytes cells containing bacteria with [Des-Thr1]-tuftsin}}{\% \text{ of phagocytes cells containing bacteria without [Des-Thr1]-tuftsin}}$

At least 800 cells had been counted.
*J. Martinez, F. Winternitz & J. Vindel, unpublished results.

in Najjar's and Fridkin's laboratories at 2×10^{-7} M (0,1 μg/ml) and a major stimulation was obtained at half that concentration (17, 20, 23). This represents a specific biological activity that rivals that obtained with several peptide hormones. It should be noted that tuftsin exerts a stimulation of long duration on the phenomenon of phagocytosis as reported by Najjar; Controls almost stop phago-cytizing after 10 mn, whereas stimulated cells continue as long as bacteria are available (6).

It is interesting to note the high potency of tuftsin in stimulating phagocytosis of *Staphylococcus aureus (Smith)* by macrophages (Table 2). The phagocytic index of controls is pretty low, due to production of a toxin, by *Staphylococcus aureus (Smith)*, which inhibits phagocytosis. Addition of Tuftsin leads to a significant increase of the phago-cytic activity of this pathogenic and toxic bacteria by mouse peritoneal macrophages. It is worth mentioning that [Des-Thr1]-tuftsin was found to have a strong inhibitory effect on phagocytosis of *Listeria monocytogenes* (Table 3) according to Fridkin *et al.* (20).

Phagocytosis kinetic had been studied with mouse peritoneal macrophages, using Van Furth's technic (24). These experiments allowed us to count total engulfed bacteria and get quantitative results. The ratio between bacteria and phagocyte is one and equal to $1-2 \times 10^7$ cells/ml. Results are summarized in Table 4 and Fig. 1. They clearly show that tuftsin stimulates phagocytosis of *Listeria monocytogenes*. The number of bacteria in the supernatant diminished as time increased. These results are comparable with those obtained previously, but they are supposed to express better the total sum of ingested bacteria.

Reduction of nitro blue tetrazolium is a well

Table 2. In vitro phagocytosis stimulation by tuftsin of *Staphylococcus aureus (Smith)* and *Staphylococcus aureus (Londres)* by macrophages*.

	Staphylococcus aureus Smith		*Staphylococcus aureus* Londres	
	Tuftsin μg/ml	I	Tuftsin μg/ml	I
Control		0.099		0.301
Tuftsin	0.003	0.108	0.003	0.349
	0.005	0.119	0.005	0.371
	0.01	0.129	0.01	0.376

$I = \dfrac{\text{number of cells containing bacteria}}{\text{total amount of scored cells}}$

At least 800 cells had been scored
*J. Martinez, F. Winternitz & J. Vindel, unpublished results.

126

Table 4. Phagocytosis stimulation by tuftsin of *Listeria Monocytogenes* by mouse peritoneal macrophages: Van Furth technic.

	Time intervals of sampling (mn)	N = Number of bacteria/ml	$N_t/N_0 \cdot 100$
Control	0	1 090 000	100
	30	1 300 000	119.3
	60	928 250	85.2
	120	1 295 125	118.8
Tuftsin 10 µg/ml	0	1 295 000	100
	30	1 423 330	109.9
	60	1 013 000	78
	120	754 625	58
Normal serum	0	1 219 330	100
	30	1 143 330	93
	60	866 250	71.1
	120	985 500	80

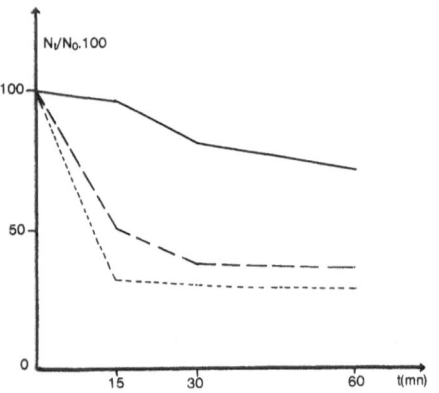

Fig. 2. Intracellular bactericidal activity by tuftsin (*Listeria monocytogenes*). (—— control. — — — tuftsin 10 mg/kg. - - - tuftsin 20 mg/kg).

known method for studying the potency of a compound to stimulate phagocytosis. Direct correlation between increased phagocytic activity of PMN leucocytes and stimulation of the cell's hexose monophosphate shunt has been well established (10, 11) and reduction of NBT is one of the assays used to evaluate this stimulation. Effects of tuftsin upon this process in human PMN leucocytes has been investigated (19, 20). The results clearly showed that tuftsin stimulated the reduction of NBT in human PMN. It was found, however, [Des-Thr[1]]-tuftsin, was a potential inhibitor of this reduction. This inhibitory effect is reversible, and the response of PMN leucocytes cells to tuftsin can be regained by washing with phosphate buffered saline. Inhibition is presumably induced by binding competition at specific sites located on the outer membrane of the cell (25).

On the other hand, tuftsin has been shown to generate a chemotactic response (26, 27) which is one of the steps in the phagocytic process.

Intracellular bactericidal activity

Materials and methods

Intracellular bactericidal activity had been studied following the technic described by Van Furth. A bacterial suspension (1 ml) of *Listeria monocytogenes* containing $1-2 \times 10^6$ bacteria/ml in gelatin Hank's solution with 10% fetal serum containing an amount of tuftsin corresponding to 0, 10 or 20 mg/kg of body weight, was injected intraperitoneally into the mouse. After five minutes, the animal is killed and a 2 ml saline solution was injected IP to harvest macrophages. The cellular suspension was collected in a glass stoppered tube freshly siliconized, at 0 °C, thoroughly washed twice with Hank's solution and centrifuged 4 mn at 110 g. The number of cells was adjusted to $6-8 \times 10^6$ cells/ml and 1 ml aliquots were collected and incubated at 37 °C. At intervals thereafter of 0, 15, 30 and 60 mn, the number of intracellular viable bacteria was determined; for this purpose, tubes were centrifuged at 110 g during 4 mn. Supernatant was discarded and 1 ml of cold distilled water containing bovine serum albumin was added. The suspension was frozen and thawed 3 times (−170, +37 °C) and appropriate dilutions were plated on petri dishes. Viable colony count was performed. Bactericidal index was expressed as $K_t = \log N_0 - \log$

N_t; N_0 represents number of intracellular viable bacteria at t_0 and N_t at time t.

Results and discussion

Bactericidal activity is enhanced by tuftsin and increased with time, (Table 5, Fig. 2). K_{60} is fairly high at 10 mg/kg of tuftsin as compared with K_{60} of the control. Comparison of bactericidal activity between t_0 and t_{15} showed that tuftsin greatly accelerate this phenomenon in its initial phase (15 mn), probably because the stimulatory effect of tuftsin does not last longer than that time interval (22, 28, 29), because of its destruction by cells enzymes (30). This stimulation seems to be concentration dependent. It is interesting to note that the t_0 number of intra-macrophages bacteria per 10^6 cells is higher for tuftsin treated animals than for controls. This confirms previous results on the role of tuftsin in phagocytosis stimulation.

Table 5. Intracellular bactericidal activity of tuftsin (*Listeria monocytogenes*).

	t	N = number of viable bacteria/10^6 cells	K_t	$N_t/N_0 \cdot 100$
Control	0	35 550		100
	15	33 725	0.023	94.8
	30	28 472	0.096	80
	60	25 045	0.152	70.4
Tuftsin 10 mg/kg I.P.	0	60 442		100
	15	30 357	0.299	50.2
	30	22 378	0.431	37.0
	60	21 900	0.441	36.2
Tuftsin 20 mg/kg I.P.	0	61 515		100
	15	19 293	0.504	31.4
	30	17 328	0.552	28.2
	60	16 666	0.567	27.1

In vivo studies*

In vivo studies of phagocytosis stimulation and bactericidal activity of tuftsin were performed through blood clearing and bacterial killing abilities of phagocytes. These activities were also studied in the liver and spleen, whose roles are paramount in these functions. Stimulation of these activities by tuftsin, was studied also in leukemic mice, and compared to control healthy mice.

Blood clearing

Material and methods

COBS (CD-1 strain) mice were used. Four types of bacteria were utilized and the appropriate concentration to be injected determined as indicated below:

Staphylococcus aureus S 57: $2.5-5 \times 10^8$
Listeria monocytogenes: $1-10 \times 10^7$
Escherichia coli: $1-5 \times 10^7$
Serratia marcescens: $1-10 \times 10^7$

Mice were injected I.V. with 0.25 ml/mouse of bacterial suspension.

Tuftsin (10 mg/kg or 20 mg/kg of body weight) was injected I.P. immediately after bacterial injection. At the indicated times, aliquots of 0.5 ml of blood were taken by cardiac puncture, with a syringe washed with isotonic NaCl solution containing 500 units of heparin/ml. Blood from 3 mice was collected, diluted and samples plated on petri dishes. After incubation at 37 °C for 24 h, the number of colonies was scored. Results are expressed per 1 ml of blood. Each experiment was repeated three times.

Results and discussion

Tuftsin stimulates blood clearing of *Staphylococcus aureus S 57* significantly (Table 6, Fig. 3). It is known that blood clearing of this bacteria is usually fast, but tuftsin treated animals showed a pronounced acceleration of this phenomenon, as compared with controls. A favorable effect of tuftsin on blood clearing of *Listeria monocytogenes* was also observed (Table 7, Fig. 4), but the virulence of this bacterium is related to its

---

* Assays were performed at 'Centre de Recherche Clin Midy', Montpellier, France.

Table 6. Tuftsin activity on blood clearing (*Staphylococcus aureus S 57*).

	Time of sampling (t)	N = number of bacteria/ml blood	$N_t/N_5 \cdot 100$
Control	5	12 000 000	100
	15	889 200	7.4
	30	582 330	4.8
	45	61 200	0.5
	60	35 940	0.3
Tuftsin 10 mg/kg	5	25 300 000	100
	15	159 810	0.6
	30	77 930	0.3
	45	51 080	0.2
	60	10 300	0.04
Tuftsin 20 mg/kg	5	10 273 330	100
	15	29 080	0.3
	30	12 330	0.12
	45	11 550	0.11
	60	10 270	0.10

Table 7. Tuftsin activity on blood clearing of *Listeria monocytogenes.*

	Time of sampling (t)	N = number of bacteria/ml blood	$N_t/N_5 \cdot 100$
Control	5	104 070	100
	15	44 980	43.2
	30	25 110	24.1
	45	8 130	7.8
	60	2 440	2.3
Tuftsin 10 mg/kg	5	107 270	100
	15	41 790	88.9
	30	11 520	10.7
	45	6 440	6.0
	60	740	0.7
Tuftsin 20 mg/kg	5	820 500	100
	15	32 970	4.0
	30	6 590	0.8
	45	4 310	0.5
	60	2 620	0.3

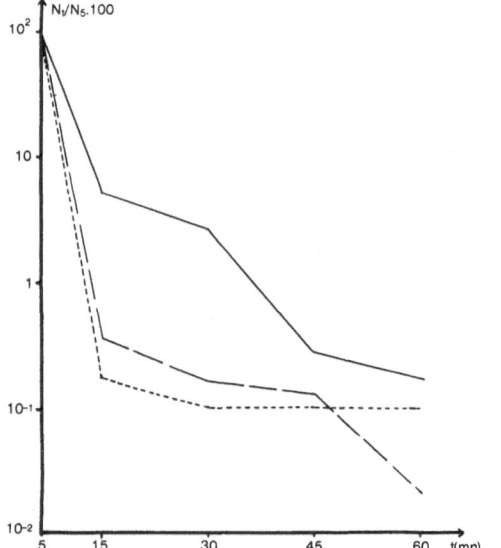

Fig. 3. Tuftsin activity on blood clearing of *Staphylococcus aureus S 57.* (— control. — — — tuftsin 10 mg/kg. - - - tuftsin 20 mg/kg).

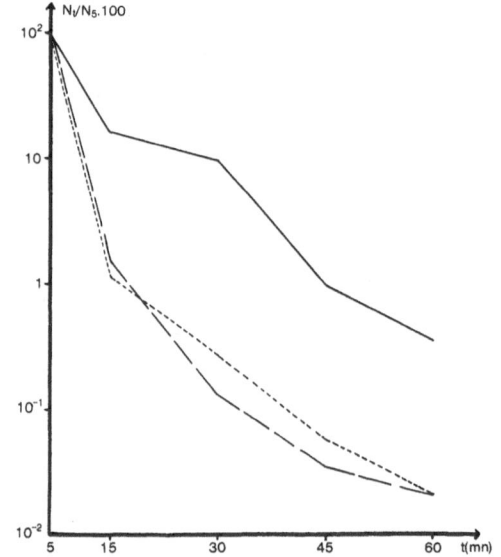

Fig. 4. Tuftsin activity on blood clearing of *Listeria monocytogenes.* (— control. — — — tuftsin 10 mg/kg. - - - tuftsin 20 mg/kg).

Table 8. Tuftsin activity on blood clearing of *Escherichia coli.*

	Time of sampling (t)	N = number of bacteria/ml blood	$N_t/N_5 \cdot 100$
Control	5	1 095 000	100
	15	338 670	30.9
	30	119 300	10.9
	45	11 410	1
	60	6 160	0.6
Tuftsin 10 mg/kg	5	1 701 750	100
	15	46 530	2.7
	30	2 920	0.2
	45	1 060	0.06
	60	610	0.04
Tuftsin 20 mg/kg	5	1 765 000	100
	15	26 450	1.5
	30	8 570	0.5
	45	1 400	0.08
	60	680	0.04

Table 9. Tuftsin activity on blood clearing of *Serratia marcescens*

	Time of sampling (t)	N = number of bacteria/ml blood	$N_t/N_5 \cdot 100$
Control	5	775 000	100
	15	80 520	10.4
	30	16 330	2.1
	45	2 280	0.3
	60	1 940	0.2
Tuftsin 10 mg/kg	5	1 696 330	100
	15	224 250	13.2
	30	5 250	0.3
	45	975	0.06
	60	830	0.05
Tuftsin 20 mg/kg	5	1 125 330	100
	15	58 430	5.2
	30	5 770	0.51
	45	1 050	0.09
	60	880	0.08

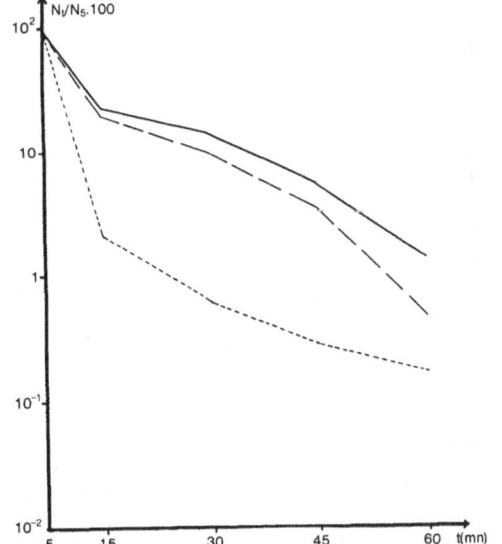

Fig. 5. Tuftsin activity on blood clearing of *Escherichia coli.* (— control. — — tuftsin 10 mg/kg. - - - tuftsin 20 mg/kg).

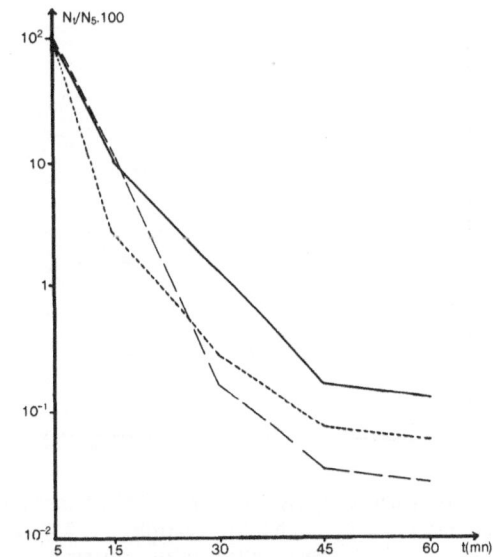

Fig. 6. Tuftsin activity on blood clearing of *Serratia marcescens.* (— control. — — — tuftsin 10 mg/kg. - - - tuftsin 20 mg/kg).

ability to resist intracellular killing by phagocytes. In fact, these bacteria are not destroyed, but they are even able to multiply inside macrophages. For *Escherichia coli,* the results on blood clearing demonstrated an interesting activity and a favorable effect (Table 8, Fig. 5). It is worthwhile to note, that there is little or no difference between the activities obtained with two concentrations of tuftsin (10 and 20 mg/kg). Similar results were observed with *Serratia marcescens* even though the blood clearing in the controls was fairly good and fast (Table 9, Fig. 6). Leukemic animals, usually show a non specific immunologic deficit against bacterial infections. Statistical studies (1975) revealed that bacterial infections are implicated in the death of more than 70% of leukemics. It was interesting to learn wether or not tuftsin was able to stimulate blood clearing of *Serratia marcescens* and *Escherichia coli* from mice with experimental leukemia. Results are summarized in Tables 10 and 11 and Fig. 7 and 8. Tuftsin stimulates blood clearing of *Serratia marcescens* and *Escherichia coli* from leukemic mice. For *Serratia marcescens,* a fairly good favorable activity was observed, but it did not reach the level of blood clearing activity of

healthy mice. For *Escherichia coli,* the beneficial effect of tuftsin is evident. It is interesting to notice that in leukemic mice, and after 60 min, there is an increase of the number of circulating bacteria, as if they already begun to multiply in the organism. Tuftsin treated mice presented a picture of blood clearing resembling that of healthy mice (Fig. 8). However, it is worth mentioning, that Constantopoulos *et al.* found that neither autologous serum, nor tuftsin were able to stimulate phagocytosis of *Staphylococcus aureus* of leukemic patients (with myelofibrosis or acute granulocytic leukemia) (15).

Bactericidal activity in liver and spleen

Phagocytes of the liver and of the spleen are relatively stationary and are strategically located to intercept invading bacteria that are circulating in the blood stream. Because of the principal role of both liver and spleen against bacterial infections, it appeared interesting to determine what happened to bacteria inside these organs, in healthy mice, in healthy tuftsin treated mice, in leukemic mice and in leukemic tuftsin treated mice.

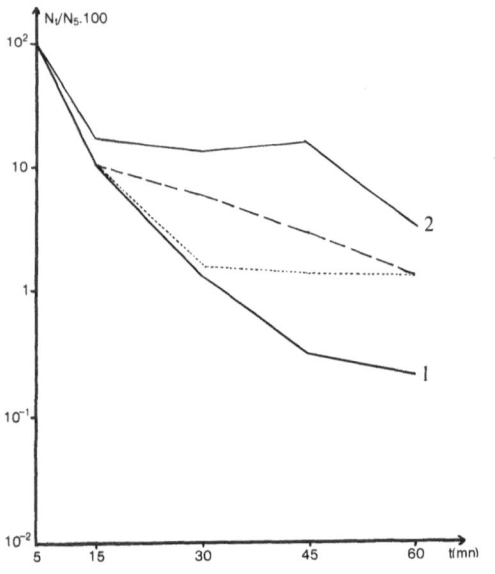

Fig. 7. Tuftsin activity on blood clearing of *Serratia marcescens* in leukemic mice. (— 1 healthy controls. — 2 leukemic controls. — — — leukemic tuftsin treated 10 mg/kg. - - - leukemic tuftsin treated 20 mg/kg).

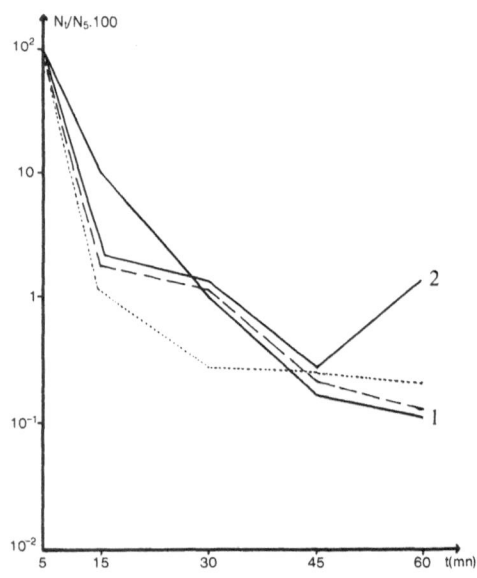

Fig. 8. Tuftsin activity on blood clearing of *Escherichia coli* in leukemic mouse. (— 1 healthy controls. — 2 leukemic controls. — — — leukemic tuftsin treated 10 mg/kg. - - - leukemic tuftsin treated 20 mg/kg).

Table 10. Tuftsine activity on blood clearing of *Serratia marcescens* in leukemic mouse.

	t of sampling	N = number of bacteria/ml blood	$N_t/N_5 \cdot 100$
Healthy controls	5	591 660	100
	15	72 500	12.25
	30	12 700	2.1
	45	3 300	0.55
	60	2 410	0.4
Leukemic controls	5	673 250	100
	15	207 360	30.8
	30	135 400	20.2
	45	140 710	20.9
	60	37 700	5.6
Leukemic + tuftsin 10 mg/kg	5	725 210	100
	15	77 600	10.7
	30	58 740	8.1
	45	38 440	5.3
	60	15 230	2.1
Leukemic + tuftsin 20 mg/kg	5	637 430	100
	15	78 400	12.3
	30	17 210	2.7
	45	13 380	2.1
	60	12 750	2

Table 11. Tuftsin activity on blood clearing of *Escherichia coli* in leukemic mouse.

	t of sampling	N = number of bacteria/ml blood	$N_t/N_5 \cdot 100$
Healthy controls	5	675 750	100
	15	77 530	11.4
	30	6 630	1.0
	45	2 170	0.3
	60	1 040	0.15
Leukemic controls	5	807 250	100
	15	33 580	4.2
	30	18 130	2.2
	45	5 060	0.5
	60	17 080	2.1
Leukemic + tuftsin 10 mg/kg	5	1 585 000	100
	15	52 280	3.3
	30	24 820	1.6
	45	6 570	0.4
	60	2 860	0.2
Leukemic + tuftsin 20 mg/kg	5	1 450 000	100
	15	23 630	1.6
	30	7 730	0.5
	45	7 340	0.5
	60	5 260	0.4

Material and methods

COBS mice (CD-1 strain) were used. The same bacteria as described above for studies of blood clearing and at same concentrations were utilized. Mice were injected I.V. with 0.25 ml/mouse of the bacterial suspension. Tuftsin (10 or 20 mg/kg) was injected I.P. immediatly after the bacteria. Mice were killed 5 and 60 min after the bacterial injection (4 mice each time).

Livers or spleens (4) were taken off and placed in

132

Table 13. Intracellular bactericidal activity in liver and spleen of mice. Effects of tuftsin (Serratia marcescens).

Organ	Tuftsin mg/kg	Time of sampling(t)	N = number of viable bacteria	N_5-N_{60}	I
Liver	0	5	8 168 700		
		60	6 843 700	1 325 000	
	10	5	7 483 300		
		60	5 485 300	2 025 000	152
	20	5	7 062 500		
		60	4 041 600	3 020 900	228
Spleen	0	5	1 392 500		
		60	679 600	712 900	
	10	5	1 715 000		
		60	771 700	943 300	132
	20	5	1 044 200		
		60	443 000	601 200	84

$$I = \frac{N_5\text{-}N_{60}\ (\text{tuftsin treated})}{N_5\text{-}N_{60}\ (\text{controls})} \times 100$$

Table 12. Intracellular bactericidal activity in liver and spleen of mice. Effects of tuftsin (Staphylococcus aureus S 57).

Organ	Tuftsin mg/kg	Time of sampling(t)	N = number of viable bacteria	N_5-N_{60}	I
Liver	0	5	117 583 300		
		60	61 733 300	55 850 000	
	10	5	128 916 700		
		60	59 500 000	69 416 700	125
	20	5	147 166 600		
		60	89 050 000	58 116 000	104
Spleen	0	5	6 435 000		
		60	4 286 600	2 148 400	
	10	5	7 045 000		
		60	3 111 600	3 933 400	183
	20	5	6 381 700		
		60	3 605 000	2 776 700	129

$$I = \frac{N_5\text{-}N_{60}\ (\text{tuftsin treated})}{N_5\text{-}N_{60}\ (\text{controls})} \times 100$$

Table 15. Intracellular bactericidal activity in liver and spleen of mice. Effects of tuftsin (*Escherichia coli*).

Organ	Tuftsin mg/kg	Time of sampling(t)	N = number of viable bacteria	N_5-N_{60}	I
	0	5	7 316 700		
		60	5 085 000	2 231 700	
Liver	10	5	8 205 000		
		60	6 175 000	2 030 000	91
	20	5	11 441 700		
		60	4 600 000	6 841 700	307
	0	5	1 055 800		
		60	776 300	279 500	
Spleen	10	5	1 143 300		
		60	556 600	586 700	210
	20	5	1 283 300		
		60	315 600	967 700	346

$$I = \frac{N_5\text{-}N_{60} \text{ (tuftsin treated)}}{N_5\text{-}N_{60} \text{ (controls)}} \times 100$$

Table 14. Intracellular bactericidal activity in liver and spleen of mice. Effects of tuftsin (*Listeria monocytogenes*).

Organ	Tuftsin mg/kg	Time of sampling(t)	N = number of viable bacteria	N_5-N_{60}	I
	0	5	16 683 300		
		60	10 470 000	6 213 300	
Liver	10	5	41 633 300		
		60	18 950 000	22 683 300	365
	20	5	57 975 000		
		60	20 350 000	37 625 000	605
	0	5	2 447 500		
		60	1 455 000	992 500	
Spleen	10	5	2 186 700		
		60	999 800	1 186 900	120
	20	5	1 252 500		
		60	397 800	854 700	86

$$I = \frac{N_5\text{-}N_{60} \text{ (tuftsin treated)}}{N_5\text{-}N_{60} \text{ (controls)}} \times 100$$

Table 16. Intracellular bactericidal activity in liver and spleen of mice. Effects of tuftsin (Serratia marcescens).

Organ	Mouse	Tuftsin mg/kg	Time of sampling (t)	N = number of viable bacteria	N_5-N_{60}	I_1	I_2
	Healthy	0	5	7 331 200	2 500 000		
			60	4 831 200			
	Leukemic	0	5	3 656 200			
			60	7 756 200			
Liver	Leukemic	10	5	5 566 200			
			60	6 417 800			
	Leukemic	20	5	6 355 700	1 137 700	45,5	
			60	5 218 000			
	Healthy	0	5	1 575 000	960 100		
			60	614 900			
	Leukemic	0	5	1 937 500	62 500		
			60	1 875 000			
Spleen	Leukemic	10	5	1 396 500	404 700	42	648
			60	991 800			
	Leukemic	20	5	1 657 200	707 600	74	1132
			60	949 600			

$$I_1 = \frac{N_5\text{-}N_{60} \text{ (leukemic treated)}}{N_5\text{-}N_{60} \text{ (healthy controls)}} \times 100 \qquad I_2 = \frac{N_5\text{-}N_{60} \text{ (leukemic treated)}}{N_5\text{-}N_{60} \text{ (leukemic controls)}} \times 100$$

tubes containing 5 ml of isotonic saline solution. The organs were homogenized and the resulting suspensions were diluted and plated on petri dishes. After incubation at 37 °C, for 60 min the number of colonies was scored.

$$\text{Bactericidal activity } I = \frac{(N_5\text{-}N_{60}) \text{ tuftsin treated}}{(N_5\text{-}N_{60}) \text{ controls}} \times 100.$$

Results and discussion

Intracellular bactericidal activity in the liver and spleen was stimulated by tuftsin as compared with the organs of untreated animals (Tables 12–15). This stimulation was particularly significant for Escherichia coli both in the liver as well as in the spleen. The ability of tuftsin to stimulate bactericidal activity of macrophages of the liver was also

Table 17. Intracellular bactericidal activity in liver and spleen of mice. Effects of tuftsin (*Escherichia coli*).

Organ	Mouse	Tuftsin mg/kg	Time of sampling (t)	N = number of viable bacteria	N_5-N_{60}	I_1	I_2
Liver	Healthy	0	5	15 116 000	4 691 000		
			60	11 425 000			
	Leukemic	0	5	11 550 000			
			60	11 626 000			
	Leukemic	10	5	10 775 000			
			60	11 150 000			
	Leukemic	20	5	10 980 000	630 000	13	
			60	10 350 000			
Spleen	Healthy	0	5	1 176 700	880 900		
			60	295 800			
	Leukemic	0	5	715 000			
			60	865 000			
	Leukemic	10	5	1 025 800	108 300	12	
			60	917 500			
	Leukemic	20	5	1 240 800	522 100	59	
			60	718 700			

$$I_1 = \frac{N_5\text{-}N_{60} \text{ (leukemic treated)}}{N_5\text{-}N_{60} \text{ (healthy controls)}} \times 100 \qquad I_2 = \frac{N_5\text{-}N_{60} \text{ (leukemic treated)}}{N_5\text{-}N_{60} \text{ (leukemic controls)}} \times 100$$

well expressed for *Listeria monocytogenes*. However, no bactericidal stimulation was found in the spleen, even for tuftsin at 20 mg/kg of treated mice. The results with *Serratia marcescens* are comparable with those obtained with *Listeria monocytogenes*. A surprising point arose from investigation of the outcome with *Staphylococcus aureus*. Bactericidal indexes were higher with 10 mg/kg than with 20 mg/kg tuftsin treated animals, in both liver and spleen. This phenomenon was observed again in the spleen with *Serratia marcescens* and *Listeria monocytogenes*. These results could be expressed as an inhibition of the reticuloendothelial system at 20 mg of tuftsin per kg of treated animals. Intracellular bactericidal activity of tuftsin was also evaluated in leukemic mice infected with *Serratia marcescens* or *Escherichia coli* (Tables 16 and 17). Leukemic animals did not show any bactericidal activity in their liver or in their spleen. The number of bacteria increased significantly from t_5 to t_{60}. Livers of tuftsin treated leukemic animals (20 mg/kg) showed some bactericidal activity as compared to leukemic controls but without reaching the values obtained in healthy animals. Tuftsin stimulation of bactericidal activity in the spleen of leukemic mice gave better results. Some beneficial effects were already obtained in 10 mg/kg tuftsin treated animals, and improved at 20 mg/kg.

Conclusion

It is evident that tuftsin stimulates the phagocytic activity of the blood granulocyte as well as the macrophage of the reticuloendothelial system. The mechanism by which tuftsin stimulates phagocytosis is still obscure, but, 'regardless of the cell, tuftsin may well stimulate one single process that relates to membrane function' (32).

From all these studies, it is clear that tuftsin in the body performs a unique function in combating infection. These studies lend encouragement to the hope that when and if tuftsin is made available to the medical profession it will prove to be a welcome addition, both bactericidal and tumoricidal, to the medical armamentarium.

References

1. Winkelstein, J. A. & Drachman, R. H., 1974. Symposium on infectious disease. Pediatric Clinics of North America 21: 551–569.
2. Fidalgo, B. V. & Najjar, V. A., 1967. Proc. Natl. Acad. Sci. U.S. 57: 957–964.
3. Fidalgo, B. V. & Najjar, V. A., 1967. Biochemistry 6: 3386–3392.
4. Najjar, V. A., Fidalgo, B. V. & Stitt, E., 1968. Biochemistry 7: 2376–2379.
5. Nishioka, K., Satoh, P. S., Constantopoulos, A. & Najjar, V. A., 1973. Biochim. Biophys. Acta 310: 230–237.
6. Najjar, V. A. & Nishioka, K., 1970. Nature 228: 672–673.
7. Nishioka, K., Constantopoulos, A., Satoh, P. S. & Najjar, V. A., 1972. Biochem. Biophys. Res. Comm. 47: 172–179.
8. Najjar, V. A., 1976. J. Pediat. 87: 1121–1124.
9. Najjar, V. A., 1976. In: Biological Membranes (Chapman and Wallach, eds.). London, Academic Press, pp. 191–240.
10. Stossel, T. P., 1974. N. Engl. J. Med. 290: 774–780.
11. Stossel, T. P., 1974. N. Engl. J. Med. 290: 833–839.
12. Baehner, R. L. & Nathan, D. G., 1968. N. Engl. J. Med. 278: 971–976.
13. Park, B. H., Figrig, S. M. & Smithwick, E. M., 1968. Lancet 2: 532–534.
14. Park, B. H. & Good, R. H., 1970. Lancet 2: 616.
15. Constantopoulos, A., Najjar, V. A. & Smith, J. W., 1972. J. Pediat. 80: 564–572.
16. Martinez, J., Winternitz, F. & Vindel, J., 1977. Eur. J. Med. Chem. 12: 511–516.
17. Najjar, V. A., 1980. Macrophages and Lymphocytes, Part A (Escobar, M. R. & Friedman, H., eds.), pp. 131–147.
18. Najjar, V. A. & Constantopoulos, A., 1972. J. Reticuloendothel. Soc. 12: 197–215.
19. Fridkin, M., Stabinsky, Y., Zakuth, V. & Spirer, Z., 1977. Peptides 1976 (Loffet, A., ed.), pp. 541–549.
20. Fridkin, M., Stabinsky, Y., Zakuth, V. & Spirer, Z., 1977. Biochim. Biophys. Acta 496: 203–211.
21. Bar-Shavit, Z., Stabinsky, Y., Fridkin, M. & Goldman, R., 1979. J. Cell. Physiol. 100: 55–62.
22. Constantopoulos, A. & Najjar, V. A., 1972. Cytobios. 6: 97–100.
23. Spirer, A., Zakuth, V., Golander, A., Bogair, N. & Fridkin, M., 1975. J. Clin. Invest. 55: 198–200.
24. Van Furth, L., 1973. Handbook Exp. Immunol. 2:
25. Nair, R. M. G., Ponce, B. & Fudenberg, H. H., 1978. Immunochemistry 15: 901–907.
26. Chaudhuri, M. K. & Najjar, V. A., 1979. Analytical. Biochem. 95: 305–310.
27. Yajima, H., Ogawa, H., Watanabe, H., Fujii, N., Kurobe, M. & Miyamoto, S., 1975. Chem. Pharm. Bull. 23: 371–374.
28. Najjar, V. A., 1974. Adv. Enzymol. 41: 129–178.
29. Nishioka, K., Constantopoulos, A., Satoh, P. S., Mitchell, W. M. & Najjar, V. A., 1973. Biochim. Biophys. Acta 310: 217–229.
30. Rauner, R. A., Schmidt, J. J. & Najjar, V. A., 1976. Mol. Cell. Biochem. 10: 77–80.
31. Hisatsune, K., Kobayashi, K., Nozaki, S. & Muramatsu, I., 1978. Microbiol. Immunol. 22: 581–584.
32. Najjar, V. A., 1978. Expl. Cell. Biol. 46: 114–126.

Received August 4, 1980.

Index to volume 41